PCR Volume 1

The Practical Approach Series

SERIES EDITORS

D. RICKWOOD
Department of Biology, University of Essex
Wivenhoe Park, Colchester, Essex CO4 3SQ, UK

B. D. HAMES
Department of Biochemistry and Molecular Biology, University of Leeds
Leeds LS2 9JT, UK

Affinity Chromatography
Anaerobic Microbiology
Animal Cell Culture
Animal Virus Pathogenesis
Antibodies I and II
Biochemical Toxicology
Biological Membranes
Biosensors
Carbohydrate Analysis
Cell Growth and Division
Cellular Calcium
Cellular Neurobiology
Centrifugation (2nd edition)
Clinical Immunology
Computers in Microbiology
Crystallization of Proteins and
    Nucleic Acids
Cytokines
Directed Mutagenesis
DNA Cloning I, II, and III
Drosophila
Electron Microscopy in Biology

Electron Microscopy in
    Molecular Biology
Essential Molecular Biology I
    and II
Fermentation
Flow Cytometry
Gel Electrophoresis of Nucleic
    Acids (2nd edition)
Gel Electrophoresis of Proteins
    (2nd edition)
Genome Analysis
HPLC of Small Molecules
HPLC of Macromolecules
Human Cytogenetics
Human Genetic Diseases
Immobilised Cells and
    Enzymes
Iodinated Density Gradient
    Media
Light Microscopy in Biology
Liposomes
Lymphocytes
Lymphokines and Interferons

PCR

Volume 1

A Practical Approach

Edited by

M. J. McPHERSON

Centre for Plant Biochemistry and Biotechnology,
Department of Biochemistry and Molecular Biology,
University of Leeds,
Leeds LS2 9JT, UK

P. QUIRKE

Department of Pathology,
University of Leeds,
Leeds LS2 9JT, UK

and

G. R. TAYLOR

Yorkshire Regional DNA Laboratory,
Belmont Grove,
Leeds LS2 9NS, UK

Oxford University Press,
Walton Street, Oxford OX2 6DP

Oxford   New York
Athens   Auckland   Bangkok   Bombay
Calcutta   Cape Town   Dar es Salaam   Delhi
Florence   Hong Kong   Istanbul   Karachi
Kuala Lumpur   Madras   Madrid   Melbourne
Mexico City   Nairobi   Paris   Singapore
Taipei   Tokyo   Toronto
and associated companies in
Berlin   Ibadan

Oxford is a trade mark of Oxford University Press

Published in the United States
by Oxford University Press Inc., New York

© Oxford University Press, 1991, except Chapter 4

First published 1991
Reprinted (with corrections) 1992 (twice), 1993 (twice), 1994, 1996

A catalogue record for this book is available from the British Library
Library of Congress Cataloging in Publication Data

Polymerase chain reaction : a practical approach / edited by
M. J. McPherson, P. Quirke, and G. R. Taylor.
p. cm.— (Practical approach series)
Includes bibliographical references and index.
1. Polymerase chain reaction. I. McPherson, M. J. II. Taylor, G. R. (Graham R.)
III. Quirke, P. (Philip) IV. Series.
(DNLM: 1. DNA Replication. 2. Gene Amplification. QH 450 P7833)
QP606.D46P66 1191   574.87'3282—dc20   90—14354
ISBN 0 19 963196 4 (Pbk)

The polymerase chain reaction (PCR) process is covered by US patents numbered 4 683 202,
4 683 195, and 4 965 188 issued to Cetus Corporation and by other issued and pending
patents worldwide

Printed in Great Britain by Information Press Ltd, Oxford, England

Preface

The polymerase chain reaction (PCR) has rapidly become established as one of the most widely used techniques of molecular biology, and with good reason; it is a rapid, inexpensive, and simple means of producing microgram amounts of DNA from minute quantities of source material and is relatively tolerant of poor quality template. Starting materials for gene analysis and manipulation by the PCR may be genomic DNA (in extreme cases from a single cell or a few microdissected chromosome fragments), RNA (perhaps from only a few cells), nucleic acids from archival specimens, cloned DNA, or PCR products themselves.

Many variations on the basic procedure have now been described and applied to a range of disciplines. In medicine, for example, the PCR has had a major impact on the diagnosis and screening of genetic diseases and cancer, the rapid detection of fastidious or slow growing microorganisms and viruses, such as mycobacteria and HIV, the detection of minimal residual disease in leukaemia, and in HLA typing. The amplification of archival and forensic material has applications in forensic pathology and evolutionary biology. PCR has established a central role in the human genome project, particularly through the concepts of sequence tagged sites, microsatellites, and interspersed repetitive sequence PCR. In most molecular biology laboratories, the PCR has found routine use in processes such as probe preparation, clone screening, mapping and subcloning, and preparation of sequencing templates, as well as for more advanced applications such as cloning very low abundance transcripts, cloning gene families, directed mutagenesis, and sophisticated gene recombination.

This volume is intended to provide a general introduction to the PCR for those new to this area, and then to cover a range of more specialized topics and applications including template preparation, gene analysis and mapping, gene cloning and manipulation, and the fidelity of DNA polymerases in PCRs. Throughout the volume there is an emphasis on practical aspects with detailed protocols forming a central feature. Occasional overlap between chapters reflects the inclusion of alternative protocols to tackle a similar problem. In a single volume of this size it is not possible to cover the complete range of applications of the PCR technology, particularly since new developments are appearing at an unprecedented rate. Nevertheless, the editors hope this volume will provide something of interest to every reader; that it will serve as a starting point for those new to the PCR yet eager to start a voyage of discovery and that it will provide a useful reference to those already well down the PCR trail.

Preface

Finally we wish to thank the authors for their valuable contributions and the staff at OUP for the speed of production of this volume.

February 1991

M. J. McPherson
P. Quirke
G. R. Taylor

Contents

ix

3. Extraction of nucleic acid from fresh and archival material 29

David P. Jackson, Jeremy D. Hayden, and Philip Quirke

4. Analysis of genomic sequence variation using amplification and mismatch detection (AMD) and direct sequencing 51

*Roland G. Roberts, A. Jane Montandon, Peter M. Green,
and David R. Bentley*

9. Inverse polymerase chain reaction 137

Jonathan Silver

10. PCR-directed cDNA libraries 147

Sarah Jane Gurr and Michael J. McPherson

11. PCR with highly degenerate primers 171

Michael J. McPherson, Kerrie M. Jones, and Sarah Jane Gurr

Contributors

TIMOTHY J. AITMAN
Nuffield Department of Surgery, University of Oxford, John Radcliffe Hospital, Headington, Oxford, OX3 9DU, UK.

DAVID R. BENTLEY
Paediatric Research Unit, Division of Medical and Molecular Genetics, UMDS, Guy's Hospital, London, SE1 9RT, UK.

C. THOMAS CASKEY
Institute for Molecular Genetics and Howard Hughes Medical Institute, Baylor College of Medicine, Houston, TX 77030, USA.

TIM CLACKSON
MRC Laboratory of Molecular Biology, Hills Road, Cambridge, CB2 2QH, UK.

RICHARD J. CORNALL
Nuffield Department of Surgery, University of Oxford, John Radcliffe Hospital, Headington, Oxford, OX3 9DU, UK.

MARGARET J. DALLMAN
Nuffield Department of Surgery, University of Oxford, John Radcliffe Hospital, Headington, Oxford, OX3 9DU, UK.

KRISTIN A. ECKERT
Laboratory of Molecular Genetics, National Institute for Environmental Health Sciences, Research Triangle Park, North Carolina 27709, USA.

PETER M. GREEN
Paediatric Research Unit, Division of Medical and Molecular Genetics, UMDS, Guy's Hospital, London, SE1 9RT, UK.

MARKUS GROMPE
Institute for Molecular Genetics, Baylor College of Medicine, Houston, TX 77030, USA.

SARAH JANE GURR
Centre for Plant Biochemistry and Biotechnology, Department of Biochemistry and Molecular Biology, University of Leeds, Leeds LS2 9JT, UK.

DETLEF GÜSSOW
MRC Laboratory of Molecular Biology, Hills Road, Cambridge, CB2 2QH, UK.

JEREMY D. HAYDEN
Department of Pathology, University of Leeds, Leeds, LS2 9JT, UK.

Contributors

CATHERINE M. HEARNE
Nuffield Department of Surgery, University of Oxford, John Radcliffe Hospital, Headington, Oxford, OX3 9DU, UK.

ADRIAN J. IVINSON
Department of Medical Genetics, St. Mary's Hospital, Hathersage Road, Manchester, M13 0JH, UK.

DAVID P. JACKSON
Department of Pathology, University of Leeds, Leeds, LS2 9JT, UK.

DANIEL H. JOHNSON
Department of Biochemistry and Molecular Biology, University of Miami School of Medicine, Miami, Florida 33136, USA.

KERRIE M. JONES
Department of Biochemistry and Molecular Biology, and Department of Genetics, University of Leeds, Leeds LS2 9JT, UK.

PETER T. JONES
MRC Laboratory of Molecular Biology, Hills Road, Cambridge, CB2 2QH, UK.

THOMAS A. KUNKEL
Laboratory of Molecular Genetics, National Institute for Environmental Health Sciences, Research Triangle Park, North Carolina 27709, USA.

SUSAN A. LEDBETTER
Institute for Molecular Genetics, Baylor College of Medicine, Houston, TX 77030, USA.

MICHAEL LITT
Department of Biochemistry and Medical Genetics, Oregon Health Sciences University, Portland, OR 97201, USA.

JENNIFER M. LOVE
Nuffield Department of Surgery, University of Oxford, John Radcliffe Hospital, Headington, Oxford, OX3 9DU, UK.

MARCIA A. McALEER
Nuffield Department of Surgery, University of Oxford, John Radcliffe Hospital, Headington, Oxford, OX3 9DU, UK.

MICHAEL J. McPHERSON
Centre for Plant Biochemistry and Biotechnology, Department of Biochemistry and Molecular Biology, University of Leeds, Leeds LS2 9JT, UK.

A. JANE MONTANDON
Paediatric Research Unit, Division of Medical and Molecular Genetics, UMDS, Guy's Hospital, London, SE1 9RT, UK.

Contributors

DAVID L. NELSON
Institute for Molecular Genetics, Baylor College of Medicine, Houston, TX 77030, USA.

ANDREW C. G. PORTER
Nuffield Department of Surgery, University of Oxford, John Radcliffe Hospital, Headington, Oxford, OX3 9DU, UK.

PHILIP QUIRKE
Department of Pathology, University of Leeds, Leeds, LS2 9JT, UK.

ROLAND G. ROBERTS
Paediatric Research Unit, Division of Medical and Molecular Genetics, UMDS, Guy's Hospital, London, SE1 9RT, UK.

BELINDA J. F. ROSSITER
Institute for Molecular Genetics, Baylor College of Medicine, Houston, TX 77030, USA.

JONATHAN SILVER
Laboratory of Molecular Microbiology, National Institute of Allergy and Infectious Diseases, National Institutes of Health, Bethesda, MD 20892, USA.

GRAHAM R. TAYLOR
Yorkshire Regional DNA Laboratory, Clarendon Wing, Belmont Grove, Leeds, LS2 9NS, UK.

JOHN A. TODD
Nuffield Department of Surgery, University of Oxford, John Radcliffe Hospital, Headington, Oxford, OX3 9DU, UK.

Abbreviations

AMD	amplification and mismatch detection
AMV	avian myeloblastosis virus
ARMS	amplification refractory mutation system
ASO	allele-specific oligonucleotide
ATP	adenosine triphosphate
BSA	bovine serum albumin
βME	β-mercaptoethanol (2-mercaptoethanol)
cDNA	complementary DNA
CEPH	Centre Etude Polymorphisme Humain
CFTR	cystic fibrosis transmembrane regulator
CIP	calf intestinal phosphatase
COP	competitive oligonucleotide priming
CTAB	cetyl trimethyl ammonium bromide
ddH$_2$O	double distilled water
dNTP	deoxyribonucleotide triphosphate
DEPC	diethylpyrocarbonate
DGGE	denaturing gradient gel electrophoresis
DMSO	dimethylsulphoxide
DTT	dithiothreitol
EDTA	ethylene diamine tetraacetic acid
EtBr	ethidium bromide
GAWTS	genome amplification with transcript sequencing
HPRNI	human placental ribonuclease inhibitor
HPRT	hypoxanthine guanine phosphoribosyltransferase
IL-2	interleukin-2
IPCR	inverse polymerase chain reaction
IPTG	isopropyl-β-D-thiogalactoside
IRS	interspersed repetitive sequences
LMP	low melting point
LTR	long terminal repeat
MES	4-morpholine-ethanol-sulphonic acid
MLV	murine leukaemia virus
mRNA	messenger RNA
nPCR	nested polymerase chain reaction
nRT	nested reverse transcriptase
OTC	orthinine transcarbamylase
PAGE	polyacrylamide gel electrophoresis
PBL	peripheral blood lymphocytes
PBS	phosphate buffered saline

PCR	polymerase chain reaction
PEG	polyethylene glycol
p.f.u.	plaque forming unit
PIC	polymorphism information content
Pipes	1,4-piperazinebis (ethane-sulphonic acid)
r.t.	room temperature
RT	reverse transcriptase
RFLP	restriction fragment length polymorphism
RNase	ribonuclease
SDS	sodium dodecyl sulphate
SSC	standard saline citrate
SOE	splicing by overlap extension
SSPE	standard saline phosphate-EDTA
SSPEn	subacute sclerosing panencephalitis
SSPR	single-strand-producing reaction
TAE	Tris-acetate-EDTA
TBE	Tris-borate-EDTA
Taq	*Thermus aquaticus*
TE	Tris-EDTA
UV	ultraviolet
VNTR	variable number of tandem repeat
VRC	vanadyl ribonucleoside complex
YAC	yeast artificial chromosome

Polymerase chain reaction: basic principles and automation

GRAHAM R. TAYLOR

1. The polymerase chain reaction

The polymerase chain reaction (PCR) is a technique for the *in vitro* amplification of specific DNA sequences by the simultaneous primer extension of complementary strands of DNA. The PCR method was devised and named by Mullis and colleagues at the Cetus Corporation (1), although the principle had been described in detail by Khorana and colleages (2, 3) over a decade earlier. The use of PCR was limited until heat-stable DNA polymerase (4) became widely available. The PCR is a major development in the analysis of DNA and RNA because it has both simplified existing technology and enabled the rapid development of new techniques which would not otherwise have been possible.

DNA polymerases carry out the synthesis of a complementary strand of DNA in the 5' to 3' direction using a single-stranded template, but starting from a double-stranded region. This is the primer extension reaction (*Figure 1*) and is the basis for a variety of labelling and sequencing techniques.

The polymerase chain reaction uses the same principle, but employs two primers, each complementary to opposite strands of the region of DNA, which have been denatured by heating. The primers are arranged so that each primer extension reaction directs the synthesis of DNA towards the other (*Figure 2*). Thus primer 'a' directs the synthesis of a strand of DNA which can then be primed by primer 'b' and *vice versa*. This results in the *de novo* synthesis of the region of DNA flanked by the two primers.

The requirements of the reaction are simple: deoxynucleotides to provide both the energy and nucleosides for the synthesis of DNA, DNA polymerase, primer, template, and buffer containing magnesium. The deoxynucleotides and primers are present in large excess, so the synthesis step can be repeated by heating the newly synthesized DNA to separate the strands and cooling to allow the primers to anneal to their complementary sequences. Initially synthesis will go beyond the sequence complementary to the other primer, but with each cycle of heating and cooling, the amount of DNA in the region

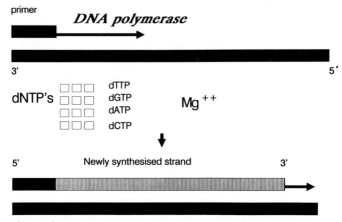

Figure 1. Primer extension. DNA polymerase extends a primer by using a complementary strand as a template.

flanked by each primer will increase almost exponentially, whilst longer sequences will only accumulate in a linear fashion, provided that the amount of starting DNA is present in limiting quantities. Thus, after several cycles the predominant reaction product will be that piece of DNA which is flanked by the primers, and will include the primers themselves.

The heating and cooling cycles can be repeated and DNA will continue to accumulate exponentially until one of the reaction products is exhausted or the enzyme is unable to synthesize new DNA quickly enough. At high DNA concentrations the DNA may also begin to prime itself and result in the synthesis of non specific products. Amplification thus either stops or produces non-specific products after a certain number of cycles. The number of cycles required for optimum amplification varies depending on the amount of starting material and the efficiency of each amplification step. Generally, 25 to 35 cycles should be sufficient to produce 100 ng–1μg of DNA of a single-copy human sequence from 50 ng of genomic DNA. A final incubation step at the extension temperature (usually 72 °C) results in fully double-stranded molecules from all nascent products.

Perhaps the most surprising feature of PCR is the stability of most of the reaction components to repeated high temperatures (denaturation is usually at 90–95 °C). Originally, Klenow polymerase was used (1), and had to be replenished after each denaturation step, but this was replaced by the heat stable *Taq* (*Thermus aquaticus*) polymerase (4–8). The use of a heat stable enzyme has two major advantages; first, replenishment after each heating step is not required, thus simplifying the process and second, the enzyme is active at higher temperatures, where annealing of the oligonucleotide primers is more specific and DNA synthesis is more rapid. Therefore the automation of PCR using heat-stable DNA polymerase is technically easier and longer, more specific reaction products can be generated (8).

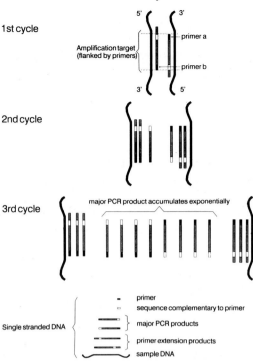

Figure 2. Schematic diagram of PCR. By using primer pairs **a** and **b** (short black lines) annealed to complementary strands of DNA (long black lines), two new strands (shaded lines) are synthesized by primer extension. If the process is repeated, both the sample DNA and the newly synthesized strands can serve as templates, leading to an exponential increase of product which has its ends defined by the positions of the primers. Products with primer at only one end (and therefore of indeterminate length) increase at a linear rate throughout the process, and together with the starting DNA form only a small fraction of the total PCR product. For clarity, the DNA and the PCR products are shown as single-stranded entities, as they would be at the denaturation stage of the reaction; the final product should be double-stranded, however, since the final step in the PCR is a synthesis step.

2. Automation of the procedure

A manual procedure for conducting PCRs has been described elsewhere in the *Practical Approach* series by one of the original investigators (9). This process is rather tedious, and rapidly creates a desire for an automated system in those who have spent more than a few days shuffling tubes between water baths, trying to keep count of the number of reaction cycles completed. Clearly, the requirements of a 'PCR machine' are extremely simple: it must repeatedly and accurately cycle the temperature of reaction vessels for a set number of counts, and a number of home-made machines have been described, as afficionados of the back pages of *Nucleic Acids Research* will be aware.

The simplest device is a robot arm fitted with a tube-rack which will move the tubes between two or three water (or oil) baths, immersing them for pre-set time periods. While such a device is effective, it requires a large amount of bench space, the temperature of the tubes is not regulated between water baths, and a fourth water bath would be required for long-term sample storage after an overnight run. For most purposes a programmable block is preferable.

Several manufacturers have produced programmable incubation blocks where the temperatures, incubation times, and cycle numbers can be selected (*Figure 3*). Although the performance of these machines is somewhat variable, most of them can be used to perform the polymerase chain reaction. Users would be well advised to obtain a K-type thermistor and compatible thermometer (preferably able to drive a chart recorder, such as Comark Electronics type 1601) to monitor the actual tube temperature as opposed to block temperature. Some manufacturers supply these with the machines. Prices of machines range from about £2000 to about £7000. Hardware and software features vary considerably between manufacturers, and careful inspection of both published specifications and actual performance of the machines is essential.

2.1 Design features of automated temperature cyclers

2.1.1 Control of temperature

Heating in most devices is by a conventional heating element which is in contact with a metal block which holds the reaction tubes. The tubes are usually 500 µl, but some machines take 1.5 ml tubes or microtitre plates. Heating rates are usually about 1 or 2°C per sec (*Figure 3*), and in some cases this ramp rate can be controlled. Not all machines provide across the block uniformity of temperature, particularly when operating on rapid cycles. Two manufacturers (Grant Instruments and Bio-med) circumvent this problem by using a heated/cooled water bath, and another (Biotherm) by using an oven which, although a little slower in adjusting the temperatures, should ensure even temperatures in all the tubes. Rapid heating rates are an advantage, since they will reduce the overall process time substantially. Methods for cooling the blocks vary between machines; the Perkin-Elmer/Cetus thermal cycler uses a conventional refrigerant which cools at about 1°C per sec whereas other machines (LEP Scientific, Coy Ltd, Pharmacia, Landgraf, and MJ Research) use solid-state cooling which cools slightly faster over the range 95–50°C. Both of these approaches mean that a reaction product can be stored at 4°C after a PCR, so that reactions can be set to run overnight. Other machines which use water (e.g. Techne) or air (e.g. Hybaid) as coolant cannot offer this facility unless connected to a separate refrigeration unit. Rapid cooling rates are an advantage since they will reduce the overall process time, although for other applications such as DNA sequencing a

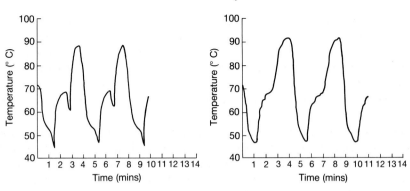

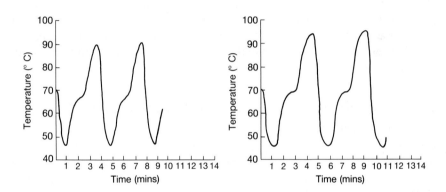

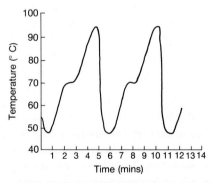

Figure 3. Temperature profiles. Internal tube temperatures monitored in several PCR devices set to cycle between 45 °C, 72 °C, and 93 °C each for 30 sec. (**a**) Three water baths with robot arm. (**b–e**) Commercially available programmable heating/cooling blocks.

5

controllable ramp rate could be useful. Most machines seem to be fairly reliable, but since they are all relatively new designs, a good service arrangement is essential.

2.1.2 Software

The basic software requirements of a programmable temperature cycler have been exceeded by most manufacturers. The minimum requirements are the ability to select three different temperatures in the range 25 to 100°C, each with an incubation time accurate to one sec in the range 1 sec to 1 h (although in most cases 5 min is adequate) and to cycle through these times up to 99 times. The machine should be able to store several of these programs. A final incubation step for longer than that used in the cycles may help the polymerase to complete the synthesis of all nascent products before analysis. This option is available on most machines. An initial denaturation period of about 5 min prior to addition of the enzyme is also useful, on some machines a separate program has to be used to do this. Other useful facilities include a hold option for extended incubation at fixed temperatures, low temperature storage, the ability to call other programs from within a program, display of the progress of the reaction, power-cut alert, elapsed time display, and the option of printer or serial output for program listings and to monitor and record the progress of a reaction. Some machines (Landgraf, Biometra) carry more than one heat/cool block which can be controlled independently, which is an advantage when optimising PCR conditions for a new set of primers.

3. Reaction components

The components of a polymerase chain reaction are readily available from commercial suppliers. A complete kit is available for Perkin-Elmer/Cetus (the 'Gene-Amp' kit) and is recommended to those who are conducting PCR for the first time, or who do not have ready access to high quality reagents. Those who wish to assemble their own reagents will require the following: DNA polymerase, deoxynucleoside triphosphates, Tris buffer, non-ionic detergent, magnesium chloride, potassium chloride, gelatin or bovine serum albumin (BSA), primers, and target DNA.

3.1 DNA polymerase

The most commonly used DNA polymerase is *Taq* polymerase isolated from *T. aquaticus* (see Chapter 14). Its advantages of heat stability and high temperature optimum make an ideal choice, although other heat-stable DNA polymerases could be used. *Taq* polymerase is available from many suppliers, including Cetus, Pharmacia, Boehringer-Mannheim, Amersham, IBI, Koch-Light, Promega, United States Biochemicals, Anglian Biotechnology, Cambio, and LEP Scientific. Alternative heat stable enzymes are becoming available, such as *Thermus thermophilus* DNA polymerase (United States Biochemicals), *Bacillus stereothermophilus* DNA polymerase (Bio-Rad), and a DNA polymerase from the extreme thermophile *Thermococcus litoralis* which is reported to possess proof-reading activity ('Vent' polymerase, New England Biolabs).

Manufacturers generally quote polymerase activity as a function of the incorporation of radiolabelled deoxynucleotides into TCA insoluble material over 30 min using nicked calf thymus DNA as a template. This means that the repair activity is being quoted, and may not always reflect the processivity of the enzyme, particularly its ability to amplify long (>1 kb) templates. Since the activity of *Taq* polymerase roughly doubles from 65°C to 72°C, the temperature of the assay is also important. *Taq* polymerase activity is optimal over a fairly broad pH range from 8.2–9.0 in 10 mM Tris (measured at 25°C), but declines at higher or lower pH.

3.2 Deoxynucleoside triphosphates (dNTPs)

Precursor dNTPs can be obtained either freeze-dried or as neutralized aqueous solutions. They are stable at −20°C for some months in either configuration, but the freeze dried reagents may require neutralization by KOH before use. Suppliers include Pharmacia, United States Biochemicals, Sigma, and Boehringer-Mannheim. When the dNTPs have been dissolved to a final concentration of 100 mM, they can be pooled in small volumes (50 µl of 20 mM of each dNTP) and stored as 100 × concentrates at −20°C for several months.

3.3 Reaction buffer

Several reaction buffer formulations have been published, but a consensus is beginning to emerge which is similar to that described by Saiki *et al.* in 1988 (8). This contains Tris at a final concentration of 10 mM (pH 8.4), 50 mM KCl, 1.5 mM $MgCl_2$, 0.01% gelatin, 0.01% NP40, and 0.01% Tween 20. The non-ionic detergents can be replaced by 0.1% Triton X-100, but some detergent is essential to obtain maximum processivity of the enzyme. It is possible to vary the salt concentrations (e.g. reduce or eliminate KCl), and some sets of primers appear to work better with raised magnesium concentrations. There also appears to be a stoichiometric interaction between the dNTPs and magnesium. Thus, higher concentrations of the dNTPs bind to magnesium and hence reduce the available magnesium concentration. Therefore, if higher concentrations of dNTPs are used it may be necessary to increase the magnesium concentration (see Chapter 14). In our laboratory, buffer is made up as required from 100 × stock solutions which are stored at room temperature apart from the dNTPs and BSA (used instead of gelatin), which are stored at −20°C.

3.4 Primers

Oligonucleotide primers are generally synthesized in the range 18–30 bases, though it is possible to amplify low complexity DNA (e.g. plasmids or previously amplified DNA) with shorter primers. The efficiency of modern DNA synthesizers using high-quality precursors is generally so high that no further purification is required after elution from the column. Primers may be conveniently stored in the ammonia eluant, which stays liquid at −20°C, enabling the dispensing of primers without repeated freeze–thawing. It is standard practice in our laboratory to prepare a working stock of mixed primers from the stocks and keep these in ammonia. Before use, 1–5 µl of this stock is heated to 95°C for 3 min to drive off the ammonia. If degradation or incomplete synthesis of the primers is suspected, they can be checked by end-labelling and analysis on a 20% denaturing gel. The results should show a predominant band with a trace of minor bands. If highly purified primers are needed (this is unusual in our experience, crude 20-mers can generally be used even for sequencing), a 20% non-denaturing polyacrylamide gel can be used to isolate the major band.

Primer sequences should have similar G+C content, minimal secondary structure (i.e. self-complementarity) and low complementarity to each other, particularly in the 3′ region. We have found that longer primers, from 24 to 30 bases work well at annealing temperatures of 60°C and over, and tend to be more robust in PCRs from difficult samples. A PCR primer selection program is available (Epicentre Software) which although fairly crude, provides some help with primer design. Perhaps DNA sequence analysis packages will develop more sophisticated programs for this purpose. The effects of redundancy and mismatches on primer efficiency are discussed elsewhere in this volume.

4. Target DNA

Many different sources have provided material for the successful amplification of DNA (see Chapters 2 and 3). The main requirements are that the DNA should be intact over the length which is to be amplified. This may mean that only shorter fragments can be amplified from sources where the DNA is degraded. It is also important to ensure that inhibitors of the reaction (e.g. detergent, EDTA, traces of phenol) are not present. In some cases a failed PCR can be recovered by using a higher dilution of the source DNA, presumably because inhibitors are also diluted out. It is essential to ensure full denaturation of the template before starting PCR, usually by heating to 93–95°C for 5 min. Since *Taq* polymerase is inactivated, albeit slowly at this temperature, it should be added after the initial denaturation step for maximum activity. This provides the additional safeguard of inactivating endogenous enzymes before adding the polymerase. When processing large batches, however, it may be more convenient to make up a large volume of buffer including the enzyme in order to reduce the number of pipetting steps and to tolerate the slight loss in enzyme activity.

5. Reaction conditions

The selection of times, temperatures, and numbers of cycles depends on the DNA being amplified and the primers chosen. Reaction volumes of between ten to 100 µl are generally used. Small volumes are an advantage for batch screening large numbers of samples, because of the savings in reagent costs, but if only a few samples are being processed it may be technically easier to work to a slightly larger reaction volume of 25–50 µl. For preparative work, such as plasmid insert amplification, 100 µl may be more useful. Concentrations of primers are generally 25–100 pmol of each primer for a 50 µl reaction, which makes many commercially supplied sequencing and reverse sequencing primers prohibitively expensive. Incubation times should be kept as short as possible, to reduce the overall cycling time and to minimize the risk of non-specific amplification. Denaturing and annealing times of 30 sec should be adequate, and extension times which allow 1 min per kilobase of target followed by a final extended incubation time (2 min per kilobase) are more than enough. The number of cycles required depends on the abundance of target and efficiency of the PCR: for the amplification of plasmid inserts from toothpicked colonies, 20 cycles is sufficient (see *Protocol 1*), for other applications more cycles may be needed. Greater specificity and sensitivity can be achieved by using nested primer pairs (see Chapters 3 and 11).

6. Detection and analysis of the reaction product

The product of a PCR should be a fragment or fragments of DNA of defined length. The simplest way to check this is to load a fraction of the reaction product and appropriate molecular-weight markers (e.g. lambda *Hin*dIII digest, or ϕX174 *Hae*III digest for small fragments) onto an agarose gel of 0.8–4% containing ethidium bromide. The product should be readily visible under ultraviolet transillumination. Small DNA products (primer-dimers) and sometimes the primers themselves can be seen as rather diffuse bands close to the leading front of the gel, but the product should appear as a sharp band at the expected size. Additional bands may be due to non-specific priming, which occurs for a variety of reasons, or to the presence of some single-stranded product, which can occur if the primers are present in unequal molar amounts. In fact the production of single-stranded DNA by the deliberate use of different primer concentrations is known as asymetric PCR (9) and can be useful for producing templates for sequencing or strand-specific probes (Section 8).

7. Preparation of probes by the PCR

The PCR can be used to circumvent the growth of bulk cultures and mini-preps of cloned DNA up to at least 6 kb. Inserts cloned into M13- and pUC-derived vectors can be amplified directly, according to *Protocol 1*, from transformed bacteria picked from ampicillin plates using M13 forward and reverse sequencing primers (*Figure 4*). Plugs taken from phage plagues are also suitable for amplification using appropriate primers.

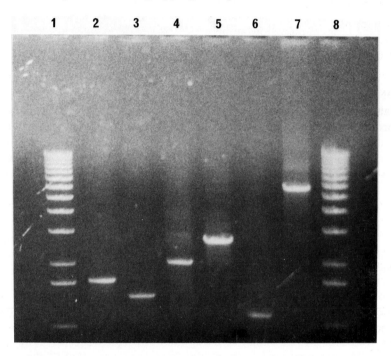

Figure 4. Amplification of plasmid inserts. Dystrophin cDNA probes cloned in BlueScript were amplified with M13 universal primer and reverse sequencing primer as described in *Protocol 1*. **Tracks 1** and **8**: 1 kb ladder (BRL), **tracks 2** to **7** amplified insets ranging from 0.8 to 6 kb.

Protocol 1. Amplification of inserts in pUC-derived vectors

1. Streak out colonies on to an LB agar plate containing 50 μg/ml ampicillin.
2. Incubate overnight at 37 °C.
3. Touch an isolated colony with a sterile toothpick and transfer the equivalent of $^1/_{10}$ to $^1/_2$ of a 2 mm colony to the bottom of a microcentrifuge tube.
4. Add 50 μl water, cover with mineral oil, and heat to 95 °C for 5 min.

10

5. Add 50 μl of 2 × PCR buffer (Section 3.3) containing 0.4 mM dNTPs, 4 units of *Taq* polymerase, and 50 pmol of each M13 primer.

M13 primers: 5'-GTTGTAAAACGACGGCCAGT-3'
 5'-AACAGCTATGACCATGATTA-3'

6. Carry out 20 cycles of [95°C, 30 sec; 60°C, 30 sec; and 72°C, 1 min per kilobase of insert]. If the insert size is not known, use a 5 min extension time. Finally, allow the samples to incubate at 72°C for the equivalent of two extension times, then cool the samples to 4°C for storage.

7. The products can be checked on agarose gels and inserts can be purified by using a low-melting-point gel, excising the band and centrifugation in a Costar Spin-X microfilter.

DNA produced in this way is suitable for use as a hybridization probe, and can be labelled by conventional means, or by the use of a modified PCR protocols to produce either double-stranded (1) or single-stranded (10, 11) probes.

8. Preparation of single-stranded DNA

Many regions of DNA can be sequenced from a double-stranded template, but more control over primer annealing conditions is possible with a single-stranded template. Single-stranded DNA can be produced by taking a small amount (1–2%) of a PCR product and repeating the temperature cycling process using one primer only for between 15 to 30 cycles (see *Protocol 2*). The single-stranded product can readily be seen on an agarose gel (*Figure 5*) and purified by electroelution followed by ethanol precipitation. Direct sequencing is then possible using the other PCR primer in standard dideoxy protocols (*Figure 6*). Alternative protocols for producing single-stranded DNA, such as the use of asymetric primers (9) or biotin-tagged primers and steptavidin-coated Dynabeads (Dynal Ltd), which can readily produce template suitable for automated sequencing (12), have also been described.

Protocol 2. Production of single-stranded DNA

1. Take 0.5–1 μl of a PCR product from a 50 μl reaction.

2. Add 50–100 pmol of one of the PCR primers.

3. Make up to 50 μl with PCR buffer (Section 3.3) containing 200 μM of each dNTP and 0.5 units of *Taq* polymerase.

4. Heat to 93°C for 5 min.

5. Carry out 15–25 cycles of denaturation (95°C for 30 sec), annealing (50–60°C for 30 sec) and extension (72°C for 1 min per kb of template).

6. Allow a final extension step at 72°C for twice the extension time in step 5.

Protocol 2. *Continued*

7. Check the product by electrophoresis on a 2% agarose gel. Two bands may be seen, a double-stranded product and a faster-running band corresponding to single-stranded DNA (see *Figure 5*).

8. Clean up the reaction product for sequencing by desalting on a spun column, e.g. Pharmacia cDNA columns, or by centrifugation in an Amicon Centricon C100 centrifugal dialysis unit. Typically, one PCR reaction should produce sufficient template for two to four sequencing reactions (see *Figure 6*).

Acknowledgements

I would like to thank Jayne Noble of the Yorkshire Regional DNA Laboratory for her valuable contribution to the work described here, and also Paul Evans of the Leeds General Infirmary Department of Haematology for the M13 forward and reverse sequencing primers.

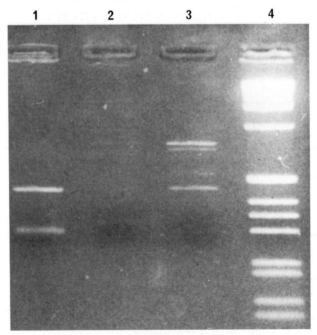

Figure 5. Single-stranded DNA from a PCR product. PCR products were extended using a single primer for 20 cycles as described in *Protocol 2*. **Track 1,** 500 bp fragment with double-stranded product (upper band) and single-stranded product. **Track 2,** 800 bp product which yielded non-specific products after 20 cycles. **Track 3,** the same source DNA after 15 cycles, with two major products corresponding to double- and single-stranded DNA. Some additional non-specific products are also visible. **Track 4** is a 1 kb ladder (BRL).

T G C A T G C A

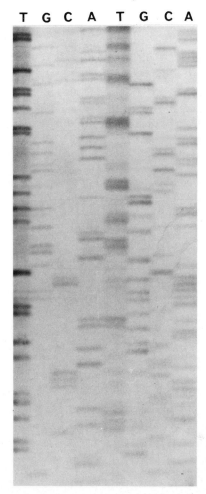

Figure 6. Direct sequencing of single-stranded DNA. The products of a single primer reaction were phenol extracted, desalted on a Pharmacia cDNA spun column, ethanol precipitated, and sequenced using an Amersham multiwell sequencing kit.

References

1. Mullis, K. and Faloona, F. (1987). In *Methods in enzymology*, Vol. 155 (ed. R. Wu), p. 335. Academic Press, New York and London.
2. Kleppe, K., Ohstuka, E., Kleppe, R., Molineux, L., Khorana, H. G. (1971). *Journal of Molecular Biology*, **56**, 341.
3. Panet, A. and Khorana, H. G. (1974) *Journal of Bioligical Chemistry*, **249**, 5213–21.
4. Chien, A., Edgar, D. B., and Trela, J. M. (1976). *Journal of Bacteriology*, **127**, 1550.

5. Powell, L. M., Wallis, S. C., Pease, R. J., Edwards, Y. H., Knott, T. J., and Scott, J. (1987). *Cell*, **50,** 831.
6. Chehab, F. F., Doherty, M., Cai, S., Kan, Y. W., Cooper, S., and Rubin, E. M. (1987). *Nature*, **329,** 293.
7. Kogan, S. C., Doherty, M., and Gitschier, J. (1987). *New England Journal of Medicine*, **317,** 985.
8. Saiki, R. K., Gelfand, D. H., Stoffel, S., Scharf, S. J., Higuchi, R., Horn, G. T., Mullis, K. B., and Erlich, H. A. (1988), **239,** 487.
9. Saiki, R. K., Gyllensten, U. B., and Erlich, H. A. (1988). In *Genome analysis: A practical approach* (ed. K. E. Davies), p. 141. IRL Press at Oxford University Press, Oxford.
10. Schowalter, D. B. and Sommer, S. S. (1989). *Analytical Biochemistry*, **177,** 90.
11. Sturzl, M. and Roth, W. K. (1990). *Analytical Biochemistry*, **185,** 164.
12. Schofield, J. P., Vaudin, M., Kettle, S., and Jones, S. C. (1989). *Nucleic Acids Research*, **17,** 9498.

2

PCR in genetic diagnosis

ADRIAN J. IVINSON and GRAHAM R. TAYLOR

1. Introduction

The PCR has revolutionized the way genetic diagnosis can be performed. Previously, the diagnosis of many genetic diseases involved taking a 10- or 20 ml blood sample followed by a lengthy protocol over a minimum of four or five days to produce an autoradiograph displaying the results. With the PCR it is possible in many cases (see *Table 1*) to perform the same test in less than 10 h starting from samples as innocuous as a mouthwash or a pinprick blood spot (1, 2). This advance is of importance to any diagnostic laboratory both in terms of streamlining the experimental aspect of the work and improving the service offered to the patient.

Table 1. Inherited disorders diagnosed using PCR protocols

Alpha-1-antitryspin	Haemophilia A and B
β thalassaemia	Huntington's chorea
Cystic fibrosis	Phenylketonuria
Duchenne muscular dystrophy	Sickle cell anaemia
Myotonic dystrophy	Familial adenomatous polyposis

In this chapter we shall (a) discuss how the PCR has increased the options available to the diagnostic laboratory, (b) describe some of the ways in which samples can rapidly be prepared and analysed, and (c) consider ways of dealing with PCR-specific problems in the analysis and interpretation of results.

2. Target DNA

2.1 Source and preparation

The starting or target DNA is one of the more forgiving reaction components and by keeping to a few simple rules problems can be avoided. Adding too much target DNA can result in non-specific amplification. Use 100–500 ng of genomic DNA as target when amplifying a single-copy sequence in 25 to 30 cycles, although anything between 100 to 1000 ng should not present a problem. The amount of target DNA should be reduced when amplifying repeat sequences. The target DNA need not be purified; however, care must be taken to minimize the presence of *Taq* polymerase inhibitors by keeping the amount of target DNA as low as possible. Preparation of tissue or cell suspensions can be limited to a simple disruption and lysing of cells and nuclei followed by centrifugation to remove the cellular debris (*Protocol 1*). Larger more dense pieces of tissue may require slicing finely followed by a proteinase K digestion step prior to a standard phenol/chloroform extraction and DNA precipitation. If using paraffin embedded tissue, remove most of the paraffin wax with a scalpel, chop the tissue finely and de-wax in xylene prior to proteinase K digestion (*Protocol 2*). Formalin-fixed tissues can yield DNA of the quality required for the PCR; however, formalin crosslinks DNA. If the exposure to formalin during the fixation has been brief, sufficient DNA may be available to serve as a target. A lengthy fixation period however may result in a very poor yield of DNA suitable to start the reaction (see Chapter 3 for further details).

Protocol 1

A. *Small pieces of soft tissue, whole blood, blood spots, sperm*

1. Finely cut approximately 10 mg of tissue or half a 5 mm blood spot or draw 30 μl of whole blood or one drop of semen and collect in a 1.5 ml Eppendorf tube.

2. Add 100 μl 0.1 M sodium hydroxide.

3. Seal tube and boil for 5 min.

4. Centrifuge ~11 000 g (microcentrifuge) for 2 min.

5. Use 5–10 μl of the supernatant for each reaction.

B. *Buccal cells*

1. Rinse mouth thoroughly with water and discard.

2. Rinse mouth vigorously for 10 sec with 10 ml sterile phosphate buffered saline (PBS) and collect in a 15 ml centrifuge tube.

3. Centrifuge for 10 min at 3000 g.

4. Discard supernatant and wash pellet in PBS[a]. A cellular pellet should be clearly visible after centrifugation. If the pelleted cells cannot be suspended in 100 μl, split them into several aliquots.

5. Proceed as in part A.

[a] PBS: 3.56 g $Na_2HPO_4.12H_2O$, 0.52 g $NaH_2PO_4. 2H_2O$, 8.5 g NaCl per litre.

Protocol 2. Hard tissues and paraffin embedded tissues

1. Finely chop approximately 50 mg of tissue and collect in a 10 ml centrifuge tube.

2. For paraffin sections remove excess wax with a scalpel and de-wax by adding 5 ml xylene and vortexing for 1 min. Pellet the tissue by brief centrifugation and discard the supernatant. Repeat and wash in ethanol.

3. Set up a standard proteinase K digestion in approximately 4 ml of salt/EDTA buffer (75 mM NaCl, 25 mM EDTA) with 200 μl of 10% SDS and proteinase K to a final concentration of 250 μl/ml. Fully digest the tissue at 37°C. This may take from 24 to 48 h. The process can be accelerated by vortexing the digest and adding more proteinase K every 12 h.

4. Phenol/chloroform extract the DNA and precipitate at −70°C for a minimum of 1 h with 2.5 vol. of ethanol and 1/20th vol. of 3 M sodium acetate. Use 500 ng DNA per reaction.

2.2 Amount of tissue and copy number

The sensitivity of the PCR is such that as little as a single copy of the target DNA is sufficient to start the amplification reaction. At the other end of the spectrum successful amplification of a highly repetitive sequence can be achieved with as much as 2 μg of starting DNA. These extremes both present some problems, and somewhere in between is an optimum. For a single-copy sequence 100–500 ng of genomic DNA is ideal. For multiple-copy sequences this is reduced to 10–100 ng. *Table 2* provides a guide to DNA yields from various tissues sufficient for between five and twenty reactions. If target DNA is in short supply it is possible to alter the PCR protocol to take this into consideration. This may involve initially increasing the annealing time to ensure a more efficient annealing process, increasing the total number of cycles, running the reaction for 10–20 cycles prior to setting up secondary

Table 2. Sources of target DNA and approximate amounts of tissues required

Source of DNA	Amount typically used	Amount of DNA
Purified genomic DNA	50–500 ng	50–500 ng
Chorionic villus samples	small frond (5 mg)	1–3 μg
Guthrie blood spot	half a 5-mm spot	0.5–1 μg
Semen	30 μl	5–10 μg
Whole blood	30 μl	0.5–1 μg
Buccal cells	one mouth wash	0.1–1 μg
Tissue blocks	50 mg	0–10 μg
Cell suspensions	5×10^5 cells	2.0–5 μg

reactions using a sample of the original reaction as the target, or using a set of nested primers to prime a new reaction after 10–20 rounds (see Chapters 3 and 11).

The great sensitivity of the PCR means that very small amounts of DNA contaminants can be amplified unwittingly. The problem of contamination and quality control is dealt with in Section 4.

3. Sex determination

3.1 Introduction

The PCR can be used to quickly establish sex from a variety of different tissues. Primers flanking part of a Y-chromosome-specific sequence (in this case the repetitive 3.4 kb sequence from the DYZI family) (3, 4) are used according to a standard PCR amplification procedure. The parameters detailed in *Protocol 3* give satisfactory results, although repetitive sequences are comparatively easy to amplify and may work well with a variety of conditions. Since testing a female sample produces a negative result it is recommended that a control is built into the test, such as a duplicate set of samples with a control set of primers. Primers flanking a ubiquitous *Alu* repeat sequence are suitable (4, 5). Those shown in *Protocol 3* give rise to a 130 bp fragment in both male and female samples. This control only ensures the integrity of the sample being tested. If the *Alu* primers are added to the Y-specific primers in the same reaction tube then the control also confirms the reaction system is working but still fails to give a good control for the Y-specific primers in that tube. For this reason control male and female samples should always be run, as should negative controls (no target DNA), to check for the presence of any contaminating target DNA (see Section 4).

Protocol 3. Sex determination by PCR

Y-specific primers: Y1 5′-TCCACTTTATTCCAGGCCTGTCC-3′
                          Y2 5′-TTGAATGGAATGGGAACGAATGG-3′

Alu control primers: A1 5′-GGCACTTTGGGAGGCCAAGG-3′
                          A2 5′-TACAAGCTTGTGCCATGCCCAAC-3′

1. Mix the following reactants:

- H_2O to   50 μl
- 10 × PCR buffer[a]   5 μl
- 2mM dNTP's   5 μl
- *Taq* polymerase   2 units
- DNA   500 ng
- primers   50 pmol of each

18

2. Denature for 5 min at 95 °C

3. Perform 20 cycles of denaturation at 95 °C (30 sec), annealing at 58 °C (30 sec) and extension at 72 °C (30 sec)

4. Final extension: 2 min at 72 °C

5. Store at 4 °C

ᵃ 10 × PCR buffer: 100 mM Tris–HCl, pH 8.8, 500 mM KCl, 15 mM MgCl$_2$, 0.1% Triton-X100.

Twenty rounds of amplification will, in male samples, give rise to a 154 bp sequence. This can be seen as a clear band on an ethidium bromide-stained 2% agarose gel after electrophoresis (see *Figure 1*). This band will be absent from female samples.

Page *et al.* (6) have demonstrated the presence of a sequence common to both the X and Y chromosomes that is characterized by X-specific and Y-specific restriction fragments. Primers covering a part of the sequence which carried a polymorphic site could be used for the purposes of sexing (7). The advantage over the above sexing protocol is that both female and male samples would give rise to an amplified band, hence problems of controls for the negative female reaction are avoided. This amplification product could be digested with an *Bam*HI giving a different pattern in males and females.

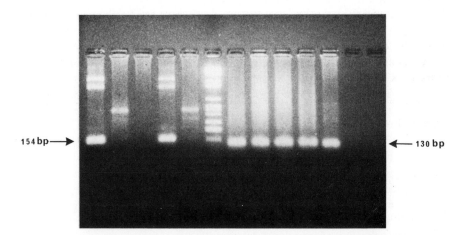

Figure 1. Sex determination. **Lanes 1–5,** Y-chromosome-specific sequence. 0.5 µg samples of genomic DNA were amplified for 20 cycles using the primers described in *Protocol 3).* One-quarter (25 µl) of each reaction was electrophoresed on an ethidium bromide-stained 2% Nusieve agarose gel. The presence of the 154 bp band (**Lanes 1** and **4**) corresponds to a male sample. **Lanes 7–11,** each sample was run in parallel with the *Alu* control primers which give rise to a 130 bp band. **Lane 6** shows a low-molecular-weight marker ladder.

19

3.2 Prenatal sexing

Prenatal sexing is often required in families with inherited sex-linked diseases. In these cases chorionic villus samples (see *Table 2*) are ideal material for fetal sexing in the first trimester of pregnancy. Fetal sexing, and indeed fetal diagnostic testing, using amniotic fluid samples, have been suggested, since the sensitivity of the PCR makes it possible to amplify sequences from the fetal cells found in the amniotic fluid. This is a relatively straightforward technique but its great sensitivity can present a problem, due to the possibility of contamination from maternal cells in the fluid. Maternal contamination of this sort does not raise problems for fetal sexing since any contaminating sequences would be female and therefore negative (however, the maternal contamination could contribute to the amplification of the control *Alu* sequence). If the fetal cells found in the amniotic fluid were to be used to screen the fetus for a deletion (for example, Duchenne muscular dystrophy prenatal diagnosis) or to examine a disease-linked restriction fragment length polymorphism (see Section 5) then maternal contamination could on occasion result in a misdiagnosis. Prenatal diagnoses of this nature should be treated with a great deal of caution.

Fetal blood sampling will provide more than enough blood for the purposes of PCR sexing, but is associated with a higher risk to the pregnancy. Lo *et al*. (8) have reported the ability to perform fetal sexing on a maternal blood sample. This advance has the great benefit of a non-invasive test for the fetus and is based on the presence of very low numbers of fetal cells in the maternal circulation. The PCR, being so sensitive, can amplify the Y-specific DNA from the fetal cells. In the case of a male fetus the result is unambiguous (given confidence in the complete absence of contamination); however, as before, PCR products from a female fetus are indistinguishable from those of the maternal DNA. Although the test could be used to screen for the presence of male fetuses the practical implications for the purposes of sexing are not clear. Should a female result be found, the diagnosis would need to be confirmed by an independent means to exclude a false-negative result. On the other hand, should a male result be found it is more than likely that further genetic analysis, looking for the presence of a specific sex-linked gene, would necessitate amniocentesis or chorionic villus biopsy. Further development of this approach requires an efficient procedure for separating fetal cells from maternal blood, which will then enable prenatal diagnosis to be carried out without the use of invasive fetal sampling.

4. Contamination

Under normal conditions the greatest risk of contamination is the possibility of introducing a small amount of previously amplified DNA into the new reaction. If, for example, some amplified sequences were carried over in a pipette or found their way into stock solutions they could act as the target DNA for that new reaction. Being specific to the primers, even a tiny contaminant of this nature could make a major contribution to the product of the new reaction. To avoid this possibility it is a wise precaution to set up new reactions using pipettes that have never been used to manipulate post-amplification products. Great care must be taken to avoid storing or manipulating post-amplification products close to PCR reagents.

Airborne contamination, such as hair roots or sloughed-off skin cells, are less of a problem in most cases. Although it is obviously best to avoid this possibility, such precautions are often not practical. A minute contribution of foreign genomic DNA to the reaction is unlikely to have a marked effect. However, if the amplification reaction is using a single cell as the source of target then even a single-cell contaminant could be critical. When undertaking such protocols special precautions must be taken (9). It has been reported that PCR primers are more resistant to ultraviolet irradiation than target DNA so that pre-exposure of a PCR mixture to UV light, for example from a germicidal lamp, can eliminate contamination (10). Using a prototype Amplirad UV source (GRI Ltd) contamination by 2×10^7 copies of a 120 bp target sequence could be completely eliminated by a 5 min exposure at a distance of 9 cm from the source with no detectable effect on primer efficiency (P. Kite, unpublished data). It may also be possible to remove contamination by pre-treating the PCR mixture with a restriction enzyme or DNase treatment that should cut the PCR product (11), although this would not remove single-stranded contaminants. In cases where PCR product contamination has become an acute problem, it can be dealt with by designing new primers which flank the original primer sites on the target sequence and hence will not amplify the original product. Of course, this would not address the problem of the source of contamination directly, and other steps will need to be taken to prevent its recurrence. Contamination problems are discussed further in Chapters 3 and 12.

5. Product analysis

5.1 Restriction fragment length polymorphism

5.1.1. Introduction

One of the most commonly used tools for following the inheritance of genes or gene markers is that of the linked restriction fragment length polymorphism (RFLP). Traditionally this technique involved the digestion of approximately 5 μg of genomic DNA, followed by electrophoresis, Southern transfer, radioactive (or, less commonly, non-radioactive) probing, and exposure of the filters to film (12). The use of the PCR has made it possible to amplify a short region of DNA surrounding the restriction site of interest which is then exposed to the relevant restriction enzyme. The amplified DNA is then visualized directly on an appropriate gel matrix. This method has many advantages over the traditional Southern analysis (see *Table 3*), not least of which is the great saving in time (1, 13).

Table 3. A comparison of Southern analysis and PCR analysis

Southern	PCR
Purified DNA	Many different tissues
5 μg Genomic DNA	0.5 μg genomic DNA
Four to five days	One day
Radioisotopes	No radioactivity

5.1.2 Primer selection and amplification

A prerequisite of the PCR-RFLP analysis is knowledge of the sequence surrounding the restriction site. The selection of primers has been discussed in general terms in Chapter 1. For the purposes of RFLP analysis it is desirable to have a reasonably short distance between primers, 100–200 bp is ideal, with the restriction site approximately central. The simplest arrangement is illustrated in *Figure 2*. Standard PCR protocols (see *Protocol 3*) apply and 20 to 30 cycles of amplification should be sufficient to produce a strong band for a single-copy sequence from 500 ng of genomic target DNA.

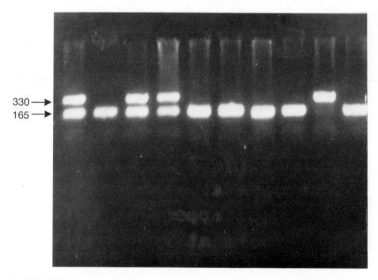

Figure 2. CS.7/*Hha* I polymorphism close to the cystic fibrosis locus of chromosome 7. 0.5 μg samples of genomic DNA were amplified for 30 cycles, digested with *Hha*I, and electrophoresed in a 2% Nusieve agarose gel. The fragments were visualized by UV fluorescence after staining with ethidium bromide. The polymorphic restriction site is central. When the site is present the 330 bp fragment is cleaved into two 165 bp fragments.

5.1.3 Restriction endonuclease digestion

In most cases it is possible to withdraw an aliquot of the amplified DNA, add the appropriate restriction enzyme, and a few hours later analyse the results on a gel. Occasionally a particular restriction enzyme may not be compatible with the PCR buffer, in which case it may be necessary to purify the amplified DNA from the reaction components prior to digestion. Purification can be a standard phenol/chloroform extraction and ethanol precipitation. Commercially available nucleic acid purification kits (Geneclean, Bio 101 Inc., or Prep-A-Gene, Bio-Rad) are ideal, and have the advantages of being faster and more efficient, reducing the amount of product lost. These purification kits are also ideally suited to purifying DNA from excised gel bands prior to sequencing. Care must be taken to avoid partial digestion which could lead to the misinterpretation of samples appearing as heterozygous for the restriction site in question (see *Figure 3*). Processing both homozygote and heterozygote controls is important to avoid this possibility. An additional confounding factor has recently been described, where a sequence variant prevented one oligonucleotide from binding to one chromosome, resulting in PCR product from only one chromosome, and thus a false result on product analysis (14). This problem ('allelic dropout') was only revealed by checking against Southern blots, and was circumvented by using a flanking primer.

5.1.4 Gel electrophoresis

DNA fragments of between 10 bp and 2.0 kb can easily be resolved on NuSieve (FMC Bioproducts) agarose gels of the appropriate concentration. A 6% gel, for example, will resolve 56 bp and 64 bp fragments (see *Figure 4*). Generally, a 2% or 3% gel will be sufficient for visualizing fragments in the 50 bp to 500 bp range if the bands are not of very similar sizes. These gels can be fragile and difficult to handle safely; however, by mixing some standard agarose (Sigma type 1) with NuSieve agarose the gels are considerably strengthened. When preparing very concentrated (4% or greater) gels the agarose should gradually be added to a vigorously stirred buffer to avoid clumping. Should finer resolution be necessary, polyacrylamide or sequencing gels can be used (see Section 5.2). Approximately a quarter of the PCR product should be sufficient to be seen directly on the ethidium bromide-stained gel. Ethidium bromide can be added to the gel to a concentration of 1 μg/ml, or gels can be stained after electrophoresis.

Running low-molecular-weight marker fragments (such as Molecular weight marker V, Boehringer-Mannheim) and previously amplified control fragments adjacent to test samples will confirm the correct size of the amplified band.

5.2 Size polymorphism

In some cases (e.g. delta F508 and I507 three base deletions in the cystic

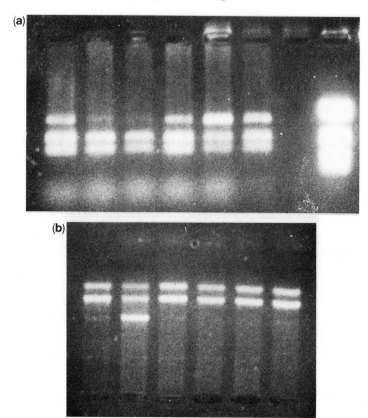

Figure 3. Partial digestion. 0.5 μg samples of genomic DNA were amplified with primers giving rise to a 380 bp band covering a polymorphic *Pst*I restriction site (in the *J3.11* locus of chromosome 7). If the site is present the 380 bp fragment is cleaved to give 230 bp and 150 bp fragments. (a) Samples were digested for 2 h at 37°C, **lanes 1, 4, 5**, and **6**, appear as heterozygotes. (b) Samples were digested for 6 h at 37°C, only **lane 5** shows up as a true heterozygote.

fibrosis transmembrane regulator (CFTR) gene, *Figure 5*) small-size poly-morphisms can be resolved on 10–12% non-denaturing (15, 16) or 8% de-naturing (17) polyacryamide gels (*Protocol 4*). This has the advantage of eliminating the use of restriction endonuclease digestion, and therefore the risk of artefacts due to incomplete digestion. The detection of microsatellite polymorphisms (see Chapter 6) and other simple repeat polymorphisms (19) also uses this approach. Native polyacrylamide gels can give rise to extra bands caused by heteroduplex formation between homologous loci. Pre-sumably the heteroduplex mismatches are retarded on the gel. These can be used to simplify carrier detection in some conditions, for example the 4 bp insertion in some Tay–Sachs cases (18), and can also distinguish between the

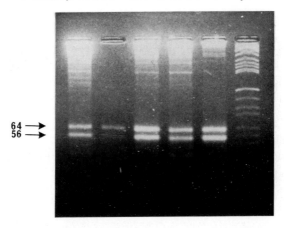

Figure 4. Electrophoretic resolution. Samples containing a 56 bp and/or a 64 bp band were electrophoresed on a 6% NuSieve agarose gel at 12 V/cm for 3 h. (Photograph courtesy of P. J. Sinnott.)

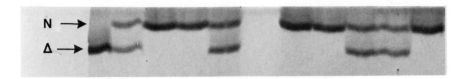

Figure 5. Resolution of small-size polymorphisms by polyacrylamide gel electrophoresis. The three base deletion in the *CFTR* gene (delta F508, the lower band) can be distinguished from the non-deleted form (upper band). **Lane 1** is homozygous for the deletion, **lane 2** is heterozygous, and **lane 3** is homozygous non-deleted.

508 and the 507 three base deletions in cystic fibrosis. Minisatellite-size polymorphisms can also be detected by the PCR, but fewer numbers of amplification steps must be performed to avoid slippage across the repeated sequences, which means that the PCR must be followed by a Southern blotting and hybridization step to detect the products (20).

Protocol 4. Use of size polymorphism to detect the delta F 508 and delta I 507 deletions in cystic fibrosis

Primers: 5'-GTTTTCCTGGATTATGCCTGGCAC-3'
            5'-GTTGGCATGCTTTGATGACGCTTC-3'

1. Mix together the following components:

- H₂O   to   50 µl
- 10 × PCR buffer (see *Protocol 3*)   5 µl
- Primers   50 pmol each

Protocol 4. *Continued*

- 2mM dNTPs   5 μl
- DNA   500 ng

For ^{35}S incorporation, reduce the nucleotide concentration of the cold dNTPs to 50 μM and add [α-^{35}S] dATP to a final concentration of 400 μM.

2. Denature at 95°C for 5 min.
3. Add 2 units of polymerase for a 50 μl reaction.
4. Perform 30 cycles at 95°C, 60°C, and 72°C, each for 30 sec.
5. Cool the sample to 4°C for storage.
6. Load 20% of the reaction product on to a non-denaturing polyacrylamide gel, or 5% of labelled product on to a sequencing gel.

If denaturing gels are used to detect the product, it must be labelled. The best resolution is obtained if ^{35}S ATP is used, either by end-labelling one primer or by including [α-^{35}S] dATP in the PCR buffer as described in Chapter 5. For analysis of unlabelled product, a 10% (29:1) polyacrylamide gel stained with ethidium bromide is satisfactory.

5.3 Other techniques

PCR products can also be characterized by allele-specific oligonucleotide (ASO) hybridization, with several advantages over the use of ASO probes for genomic hybridization. Since the PCR product is present in high concentration, the DNA is of low complexity and shorter probes with higher discrimination between normal and mutant alleles can be used. Other methods which can be used for genetic diagnosis, such as amplification mismatch detection and direct sequencing, are described in Chapters 1, 3, 4, and 5.

Acknowledgements

We thank Dr Peter Kite of the Leeds General Infirmary Department of Microbiology for information on the ultraviolet light inactivation of PCR contaminants.

References

1. Williams, C., Williamson, R., Coulette, C., Loeffler, F., Smith, J., and Ivinson, A. (1988). *Lancet*, **i,** 102.
2. Rubin, E. M., Andrews, K. A., and Waikan, Y. (188). *Human Genetics*, **82,** 134.
3. Nakahori, Y., Mitani, K., Yamada, M., and Nakagome, Y. (1986). *Nucleic Acids Research*, **14,** 7569.

4. Kogan, S. C., Doherty, M., and Gitschier, J. (1987). *New England Journal of Medicine*, **317,** 985.
5. Deininger, P., Jolly, D., Rubin, C., Friedman, T., and Schmid, C. (1981). *Journal of Molecular Biology*, **151,** 17.
6. Page, D., Mosher, R., Simpson, E., Fisher, E., Mardon, G., Pollack, J., McGillivray, B., de la Chapelle, A., and Brown, L. (1987). *Cell*, **51,** 1091.
7. Palmer, M. S., Berta, P., Sinclair, A. H., Pym, B., and Goodfellow, P. N. (1990). *Proceedings of the National Academy of Sciences, USA*, **87,** 1681.
8. Lo, Y-M., Patel, P., Wainscoat, J., Sampietro, M., Gillmer, M., and Fleming, K. (1989). *Lancet*, **i,** 1363.
9. Kitchin, P. A., Szotyori, Z., Frmholc, C., and Almond, N. (1990). *Nature*, **344,** 201.
10. Sarkar, G. and Sommer, S. S. (1990). *Nature*, **343,** 27.
11. Furrer, B., Candrian, U., Wieland, P., and Luthy, J. (1990). *Nature*, **346,** 324.
12. Maniatis, T., Fritsch, E., and Sambrook, J. (ed.) (1982). *Molecular Cloning: A Laboratory Manual*. Cold Spring Harbor Laboratory Press, Cold Spring Harbor, NY.
13. Embury, S., Scharf, S., Saiki, K., Ghalson, M., Golbus, M., Arnheim, N., and Erlich, H. (1987). *New England Journal of Medicine*, **316,** 656.
14. Fujimara, F. K., Northrup, H., Beaudet, A. L., and O'Brien, W. E. (1990). *New England Journal of Medicine*, **322,** 61.
15. Scheffer, H., Verlind, E., Penninga, D., Ter Meerman, G., Ten Kate, L., and Buys C. (1989). *Lancet*, **ii,** 1345.
16. Mathew, C. G., Roberts, R. G., Harris, A., Bentley, D. R., and Bobrow, M. (1989). *Lancet*, **ii,** 1346.
17. Taylor, G. R., Noble, J.S., Hall, J. L., Quirke, P., Stewart, A. D., and Mueller, R. F. (1989). *Lancet*, **ii,** 1345.
18. Triggs-Raine, B. L. and Gravel, R. A. (1990). *American Journal of Human Genetics* **46,** 183.
19. Ludwig, E. H., Friedl, W., and McCarthy, B. J. (1989). *American Journal of Human Genetics*, **458,** 20.
20. Jeffreys, A. J., Wilson, V., and Keyte, J. (1988). *Nucleic Acids Research*, **16,** 10953.

3

Extraction of nucleic acid from fresh and archival material

DAVID P. JACKSON, JEREMY D. HAYDEN, and PHIL QUIRKE

1. Introduction

The molecular biological study of fresh clinical material is relatively straight-forward, since good quality DNA or RNA can be easily extracted. However, the nucleic acid from archival material tends to be impure, of low molecular weight, containing single-stranded nicks, and hence is not as amenable to analysis. The sensitivity of the polymerase chain reaction (PCR) allows small quantities of poor quality genetic material to be tolerated. As long as there is sufficient single-stranded DNA or RNA to bridge the distance between the 5′ ends of the two oligonucleotide primers, then amplification can take place (1). Hence, the PCR could be used for the retrospective molecular biological study of a wide variety of pathological archival material. This also has implications for the extraction procedures that may be used to provide template DNA or RNA for the PCR, since very simple, rapid techniques should provide suitable quality template material.

2. Analysis of extracted DNA samples

When extracting DNA for use as PCR template it is very useful to know the concentration and quality of the DNA in the sample. Samples can be quantitatively assayed by spectrophotometry at 260 nm or by fluorometry. We routinely use the TKO-100 Dedicated mini-fluorometer (Hoefer Scientific Instruments, San Francisco) which measures the fluorescence enhancement of a dye, Hoescht 33258, as it binds to DNA. This is an economical and convenient instrument.

Qualitatively, the DNA can be assayed by running on a 0.7% agarose electrophoresis 'mini-gel'. A λ *Hin*dIII digest run alongside the samples gives a ladder of fragments ranging from 23 130 bp to 125 bp, allowing the molecular weight of the DNA to be estimated.

Such information allows the success of the extraction techniques to be determined, so that a negative result on PCR might be explained by a failure of the DNA extraction procedure.

3. Extraction of DNA from fresh tissue

3.1 Proteinase K incubation

Techniques for the extraction of DNA from fresh tissue, cells, or cultured cell lines are well-described in laboratory protocol manuals (2). These tend to be long and laborious but do produce good quality, high-molecular-weight DNA suitable for conventional molecular biological analysis. Such procedures involve incubating tissue with the proteolytic enzyme proteinase K for up to 24 h, followed by purification of the DNA by a number of organic extraction steps using phenol and chloroform. The DNA is then re-dissolved in a buffer and stored at 4°C.

The technique described in *Protocol 1* has several advantages. First, it is very efficient, and can produce relatively large amounts of very-high-molecular-weight DNA from as little as a single fresh frozen tissue section (approximately 10 mg) (*Figure 1*). Second, the DNA can be stored for up to several months. So this extraction method could be used to produce samples on which many analyses could be performed over a long time period.

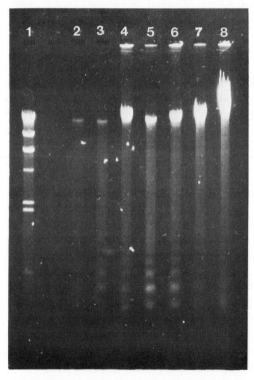

Figure 1. DNA extracted from different masses of fresh tonsil using *Protocol 1A*. **Lane 1** λ *Hin*dIII digest; **lane 2**: 10 mg; **lane 3**: 20 mg; **lane 4**: 50 mg; **lane 5**: 100 mg; **lane 6**: 200 mg; **lane 7**: 500 mg; **lane 8**: 1 g.

Most protocols suggest using a Tris–EDTA buffer for DNA storage. However, the EDTA in such buffers will chelate the Mg^{2+} in the PCR buffer, the concentration of which is vital, and so may cause problems with the sensitivity and specificity of the reaction. For this reason, we suggest simply storing the DNA in distilled water. If the DNA is dissolved in an appropriate volume of water to produce a concentration of between 100 and 500 ng/µl, such that an aliquot of 1 µl can be used as the PCR template, then even a single tissue section will produce sufficient DNA for several amplification reactions.

Protocol 1. Nucleic acid extraction by proteinase K incubation

1. A. *Fresh tissue* (Section 3.1)
   Snap freeze blocks of fresh tissue in liquid nitrogen.

   either: under liquid nitrogen, crush tissue to a fine powder using a mortar and pestle. Transfer powder to a microfuge tube.

   or: cut cryostat sections from fresh frozen tissue block on to glass slides. Using a scalpel blade, scrape sections into a microcentrifuge tube.

   B. *Fresh cytological preparations*
   Wash freshly exfoliated cells or cultured cell lines in PBS.[a] Pellet cells by centrifugation.
   Repeat.

   C. *Paraffin-embedded tissue*
   Cut 4 or 5 µm tissue sections from the paraffin-embedded block. Add directly to a microfuge tube, without de-waxing.

   D. *Haematoxylin and eosin (H&E) stained sections or archival exfoliative cytology preparations*
   Soak slide in xylene (24–48 h for H&E slides, 3–5 h for cytology slides). With a scalpel blade, prise off the coverslips. Scrape section or cells into a microfuge tube.

2. Suspend the tissue in digestion buffer[b] using 1.2 ml/100 mg tissue ($\simeq$ 25 µl/ tissue section). Incubate on a tumbler (Luckman cell mixer CM100) for 1 h at 37°C for fresh preparations, for up to 5 days for archival preparations.

3. Add an equal vol. of phenol/chloroform/isoamyl alcohol (25:24:1). Mix well. Separate phases by centrifugation. Collect upper aqueous phase using a wide-bore pipette. Repeat.

4. Add an equal vol. of chloroform/isoamyl alcohol (24:1). Mix well. Separate phases by centrifugation. Collect upper aqueous phase using a wide-bore pipette.

5. Add 2 vol. of cold absolute ethanol and one-tenth vol. of 3 M sodium acetate (pH 5.2). Allow DNA to precipitate at −20°C for at least 1 h.
   (For RNA extraction add 1 µl/ml of solution of glycogen as a carrying agent.)

Protocol 1. *Continued*

6. Pellet DNA in a microcentrifuge ($13\,000\,g$) for 10 min. Pour off ethanol.

7. Dry DNA pellet in a vacuum desiccator for 30 min.

8. Resuspend DNA in distilled H_2O, for at least 24 h at r.t.

9. Use 1 μl aliquots for PCR.

> [a] PBS: 3.56 g $Na_2HPO_4.12H_2O$, O.52 g $NaH_2\,PO_4.\,2H_2O$, 8.5 g NaCl per litre.
> [b] Digestion buffer: 100 mM NaCl, 10 mM Tris–Cl, 25 mM EDTA, 0.5% SDS pH 8, 0.1 mg/ml proteinase K (labile therefore must be added fresh).

Fresh tissue need only be incubated with proteinase K for 1 h to produce as good yields of high-molecular-weight DNA as much longer incubations, thus reducing the time required for this extraction procedure (*Protocol 1*). This method can be applied to blocks of fresh tissue such as surgical or unprocessed cadaveric specimens, cultured cells, and fresh exfoliative cytological preparations including cervical smears. The DNA produced by this method is of a suitable quality for use with standard molecular biological techniques as well as PCR.

3.2 Boiling

DNA suitable for PCR amplification can also be produced from fresh tissue and cytology specimens by simple boiling in distilled water (3) (*Protocol 2*). The major advantage of this technique is its speed, taking only a matter of minutes. Prolonged boiling of tissues reduces the yield of released DNA, so the optimum boiling time is only 15 min. Again, a small amount of tissue will produce sufficient DNA for several amplification reactions. DNA extracted by this method can be stored at 4°C or frozen, although our experience suggests not as successfully at 4°C as a purified DNA sample.

Protocol 2. DNA extraction by boiling (Section 3.2)

Fresh tissue and cytological preparations but can also be used for archival preparations when large amounts of tissue are available or when amplifying high-copy-number sequences

1. Prepare tissue or cell pellets as in *Protocol 1*.

2. Resuspend in distilled H_2O. Pierce hole in microcentrifuge tube lid.

3. Boil for 15 min (for example in boiling water bath, or in a programmable heating block).

4. Use 1 μl aliquots of boiled solution for PCR.

Boiling tissue certainly produces a DNA sample of sufficient concentration and quality to allow the amplification of a single-copy gene, although it may not always be efficient enough for the amplification and detection of lower-copy-number sequences such as persistent viral infections. In such situations, we suggest the use of the proteinase K extraction method.

4. Extraction of DNA from archival material

A wide range of different tissue types and preparations are stored in pathological archives and museums. The most common of these is routinely processed, formaldehyde fixed, paraffin-embedded tissue, but will also include haematoxylin and eosin-stained tissue sections, stained slides of exfoliative cytology samples including Papanicalaou stained cervical smears, and gross museum specimens. The ability to extract DNA from such material would allow the retrospective analysis of a range of disease processes.

4.1 Formaldehyde-fixed, paraffin-embedded material

4.1.1 Effects of processing on DNA extraction

Formaldehyde, as a 4% aqueous solution (formalin), is the most commonly used fixative in routine histopathological practice. It is inexpensive, readily available, stable, usable with almost any tissue, and very importantly, it is a non-coagulative fixative. Formaldehyde acts by forming crosslinks between protein molecules. The reaction occurs with the basic amino acid lysine, specifically the 40% to 60% of lysyl residues present on the exterior surface of protein molecules. This crosslinking of proteins forms a gel, maintaining the *in vivo* relationship of intracellular components. Soluble proteins are fixed to structural proteins and thus rendered insoluble, and the whole structure is given some mechanical strength.

However, formaldehyde is not a good fixative of nucleic acids, and DNA in its native state does not react to any extent with formaldehyde. For the reaction to occur, it is necessary to destroy the hydrogen bonds that hold together the two strands of DNA, for example by heating. This allows the reaction of formaldehyde with the amino groups on the bases now exposed by the uncoiling of the double-stranded DNA molecule. Although this reaction causes no damage to the main phosphodiester chain, the bound formaldehyde opposes any renaturation of the DNA that may occur on cooling. At temperatures usually used for fixation (i.e. 20–22 °C) no uncoiling of DNA occurs and there is no reaction of the formaldehyde with the DNA bases. Only at elevated temperatures, for example those used when tissues are impregnated with paraffin-wax during processing, might a reaction with any remaining fixative take place, thus allowing cross-linking to proteins and fixation of the DNA.

The poor fixation of DNA by formaldehyde at room temperatures, causes loss of DNA into the fixative solution by leaching. Such leaching obviously increases with the length of time that the tissue spends in fixative, and a prolonged period of fixation can lead to a dramatic reduction in the yields of DNA that can be extracted from the tissue (4). Fortunately, in most laboratories tissue for histology is routinely fixed for 24 h in formalin, causing comparatively small losses of extractable DNA. However, comparing DNA samples extracted from fresh and formalin-fixed tissue, it can be seen that even only 24 h of fixation still has a profound effect on both the quantity and quality of the DNA that can be extracted, while a delay in fixation has no appreciable adverse effect on DNA extraction. Tissue stored in screw-top containers at room temperature for up to seven days prior to fixation in formalin produce as good yields of DNA as tissue fixed immediately after surgical excision (4).

4.1.2 Extraction of DNA from paraffin-embedded material

Most recent reports describing the extraction of DNA from paraffin-embedded material involve a modification of the fresh tissue proteinase K incubation extraction technique, and are capable of producing good yields of relatively high molecular weight DNA. We and others (5) have found that increasing the length of the proteinase K incubation from 24 h up to 5 days results in up to a tenfold increase in the yield of extracted DNA with a molecular weight of up to 5 kb (*Figure 2*). This is a very efficient procedure, and will extract DNA from as little as a single paraffin-embedded tissue section. Dewaxing the tissue prior to extraction appears to be unnecessary and causes no reduction in yield (*Figure 2*). The details of the technique are given in *Protocol 1*.

Others have shown that the DNA extracted by this method is suitable for PCR amplification as well as other molecular biological techniques such as Southern blotting (6). These DNA samples are often of very high concentration but paradoxically they tend to produce weaker PCR signals than if the samples are diluted before being used as template. This phenomenon is not seen if high concentrations of DNA extracted from fresh tissue are used as PCR template, and so this inhibition of PCR may be due to a factor being co-extracted from the fixed, embedded tissue as has been previously suggested (7).

Weak or absent PCR product signals may be due to inhibition of the *Taq* polymerase by contaminants in the DNA sample, or alternatively because the template material comprises poor quality, damaged DNA. This may be determined by the presence or absence of the 'primer-dimer' bands (8). These are produced for two reasons: first, primer complementarity, and second, because *Taq* polymerase tends to add on a number of non-specific nucleotides to the ends of unused primers which may then overlap and anneal. The two primers are extended along each other by the enzyme to produce a double-stranded fragment of DNA, 40–60 bp long (depending on the length of the original primer sequences), which will appear on the PCR gel. Primer-dimers

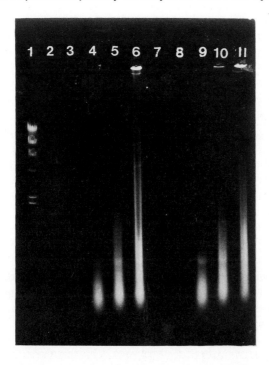

Figure 2. DNA extracted from waxed or de-waxed sections of paraffin-embedded tonsil, using different times of proteinase K incubation. **Lanes 2–6** waxed sections, and **Lanes 7–11** de-waxed sections. **Lane 1**: λ *Hind*III digest; **lane 2**: 1 h at 50°C; **lane 3**: 1 h at 50°C, 2 h at 37°C; **lane 4**: 1 h at 50°C, 24 h at 37°C; **lane 5**: 1 h at 50°C, 48 h at 37°C; **lane 6**: 1 h at 50°C, 120 h at 37°C; **lane 7**: 1 h at 50°C; **lane 8**: 1 h at 50°C, 2 h at 37°C; **lane 9**: 1 h at 50°C, 24 h at 37°C; **lane 10**: 1 h at 50°C, 48 h at 37°C; **lane 11**: 1 h at 50°C, 120 h at 37°C. (Reproduced by kind permission of the Editors of the *Journal of Clinical Pathology*.)

can provide a useful indicator of *Taq* polymerase function. Although they do not always occur there are many cases in which PCR generates primer-dimers due to limited complementarity of primers. Annealing of primers allows *Taq* mediated replication of the dimers. Clearly, if a particular primer pair are known to produce dimers their appearance is a good diagnostic tool for *Taq* polymerase function.

If the enzyme does appear to have been inhibited, the signal strength may be improved by first diluting the template DNA sample in an attempt to reduce the concentration of contaminating inhibitors. Second, the DNA template may be further purified using commercial kits designed for use with small quantities of DNA, and we have found the Isogene kit (ILS Ltd, London), used according to the manufacturer's directions, to be of some value. Further improvements in signal strength may be produced by increasing the concentration of the *Taq* polymerase or altering the concentration of Mg^{2+} in the PCR buffer.

However, if primer-dimers do appear on the PCR gel, and the specific product signal is weak or absent, then this is likely to be due to the poor quality of the template. This can be identified prior to starting a PCR for an unknown sequence by the use of a genomic positive control such as Factor VIII which should be amplifiable on every occasion, when using human DNA. If this fails and primer-dimers are observed then poor quality template is probably the problem. In this situation, the product may be improved by increasing the concentration of the template DNA; or alternatively, increasing the annealing and extension times during early cycles of PCR or resorting to nested PCR (nPCR). A balance must therefore be struck between having sufficient good quality DNA template in the reaction tube whilst keeping the concentration of contaminating inhibitors to a minimum (*Figure 3*).

Owing to its reduced requirements in terms of template DNA quantity and quality, the PCR will tolerate DNA extracted from paraffin-embedded tissue using a shorter proteinase K incubation. Even when paraffin-embedded tissue is boiled in distilled water (*Protocol 2*), an aliquot can be used as PCR template, allowing the successful amplification of a single-copy gene such as Factor VIII. Such a technique could have a place for the rapid amplification of high-copy-number targets, but its reliability in our hands for low-copy-number sequences is limited.

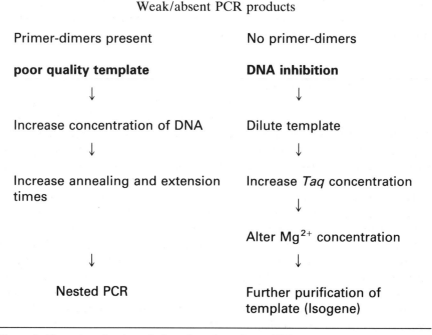

Weak/absent PCR products

Primer-dimers present	No primer-dimers
poor quality template	**DNA inhibition**
↓	↓
Increase concentration of DNA	Dilute template
↓	↓
Increase annealing and extension times	Increase *Taq* concentration
	↓
	Alter Mg^{2+} concentration
↓	↓
Nested PCR	Further purification of template (Isogene)

Figure 3. Flow chart for the solution of weak or absent PCR products.

4.1.3 Extraction of DNA from haematoxylin and eosin (H&E) stained sections

Haematoxylin and eosin (H&E) stained sections are prepared from nearly all blocks of tissue processed for histological examination, and hence are readily available. These sections are usually 4 or 5 μm thick, cut from the paraffin-embedded tissue block, de-waxed, stained, and then mounted on a slide in resin under a glass coverslip. Originally the resin used was the natural substance Canada balsam, but more recently this has been replaced by synthetic styrene-based resins. Both, however, are soluble in xylene.

DNA can be extracted from H&E sections by first soaking the slide in xylene for 24–48 h and then using a scalpel blade to prise off the coverslip. The stained section can then be scraped into a microfuge tube and incubated with proteinase K for up to 5 days. This extraction can be performed on a single H&E-stained section up to 30 years old, producing very good yields of DNA. In fact, greater yields of DNA can be extracted from archival H&E-stained sections than unstained sections freshly cut from the same paraffin-embedded tissue block (9) (*Figure 4*). This is probably due to the fact that resin-mounted, covered sections are less exposed to the atmosphere than tissue in a paraffin-embedded block, but may be related to an interaction by the paraffin-wax or the mountant with the DNA in the tissue.

The disadvantage of using H&E-stained sections as a source of DNA, is that the tissue is obviously destroyed during the extraction procedure. However, archival sections may prove valuable; for example, when the original tissue block may have been lost.

4.2 Effect of fixative agent on DNA extraction

While an aqueous ('tap water') solution of formaldehyde is the most commonly used fixative, there are a number of other agents which can be utilized. The choice of fixative depends on its properties and the technique to be applied to analyse the tissue. For example, certain specialized investigations such as electron microscopy, histochemistry, and immunofluorescence require specialized fixation techniques.

The agent in which tissue has been fixed will have a profound effect on the DNA within the tissue, and on the ability of the DNA to be extracted. Different fixatives vary greatly in this respect.

Using a proteinase K incubation technique we have shown that DNA can be extracted from tissue fixed in a 'tap water' solution of formaldehyde, although the yield and molecular weight of the DNA are much lower than from fresh tissue. It is also possible to extract DNA from tissue fixed in other fixative agents, such as neutral-buffered formalin, paraformaldehyde and even formol sublimate (which contains mercuric chloride). The yields and quality of this DNA are similar to those for formalin-fixed tissue. Bouin's

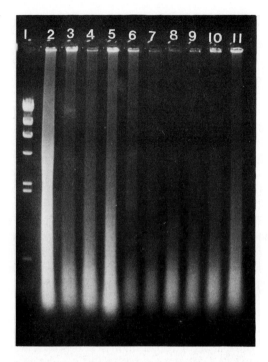

Figure 4. 0.7% agarose electrophoresis gel showing DNA extracted from archival Papanicalaou cervical cytology smear from 1969 **(lane 2)**. Paraffin-embedded sections from 1969 **(lane 3)**; 1979 **(lane 6)** and 1984 **(lane 9)**. Freshly cut, H & E stained sections from 1969 **(lane 4)**, 1979 **(lane 7)** and 1984 **(lane 10)**. Archival H & E stained sections from 1969 **(lane 5)**, 1979 **(lane 8)** and 1984 **(lane 11)**. Lane 1 is a λ *Hind*III digest standard ladder. (Reproduced by permission of the Editor of *Nucleic Acids Research*.)

reagent is a picric-acid-based fixative which has an adverse effect on DNA in the tissue. However, it is possible to extract DNA from tissue fixed in this reagent, although the yields of DNA are reduced when compared to formalin. In contrast, however, tissue fixed in Carnoy's reagent (containing ethanol, chloroform, and acetic acid), when incubated with proteinase K, produces very good yields of high-molecular-weight DNA, including some fragments larger than 9 kb (*Figure 5*). DNA samples extracted from tissue fixed in all these different fixative agents were suitable for amplification of a single-copy gene using the PCR.

Hence it seems that although the process of fixation of tissue has a profound effect on the DNA that can be extracted from it, DNA suitable for PCR can still be extracted from tissue fixed in a variety of different fixatives, allowing a wider range of fixed tissue to be used for molecular analysis. Also, although Carnoy's reagent is not ideal as a fixative for histological techniques, it can be used to fix tissue very rapidly whilst preserving relatively high quality

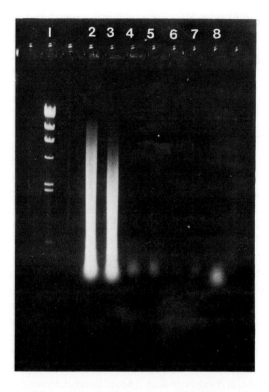

Figure 5. DNA extracted from 100-mg blocks of tonsil fixed in a range of fixative agents. **Lane 1**: λ *Hind*III digest; **lane 2**: Carnoy's reagent (5 h); **lane 3**: Carnoy's reagent (24 h); **lane 4**: Formalin (24 h); **lane 5**: Neutral buffered formalin (24 h); **lane 6**: Formol sublimate (24 h); **lane 7**: Paraformaldehyde (24 h); **lane 8**: Bouin's reagent (24 h).

DNA (4). Hence this or similar reagents may be useful for the storage of tissue specimens where low temperature storage for fresh tissue is not available.

Recently, the rapid fixation of tissue, by microwave irradiation of the specimen in the fixative solution, has been reported. This technique, requiring only several minutes of fixation time, obviously has a number of advantages for the histologist; however, tissue fixed by this method is still amenable to DNA extraction by proteinase K incubation, and this DNA is again suitable for PCR amplification (P. Jackson, unpublished data).

4.3 Exfoliative cytology specimens

Exfoliative cytology specimens are those collected by washing, scraping, or aspiration techniques, and include cells found in urine, sputum, pleural effusions, and most notably cells from the uterine cervix. Once obtained, the cells are placed on a glass slide, fixed, stained, and mounted under a coverslip. Such archival slides, particularly cervical smears, potentially provide a valuable and easily obtainable source of both normal and abnormal cytological material, but have not previously been utilized for molecular biological techniques. The extraction of DNA from archival exfoliative cytology specimens has recently been described (9). This can be performed using the same extraction method as applied to the H&E-stained sections (*Protocol 1*): removing the coverslip by soaking in xylene, scraping the cells into a microfuge tube and incubating with proteinase K at 37°C, for 3–5 days, before purifying the DNA by organic extraction. We have found that archival cytology preparations only require soaking in xylene for 3 to 5 h to remove the coverslip. Cytological preparations such as those from urine samples or pleural effusions contain only very small numbers of cells, too few to be successfully scraped off the slide with a scalpel. However, DNA can still be extracted from these specimens by painting nail varnish around the area of cells on the slide to form a well. This well can be filled with the proteinase K digestion buffer and the whole slide then incubated. The solution can be removed from the well using a pipette, and purified by organic extraction.

Since such preparations contain so few cells, they only produce very small yields of low-molecular-weight DNA, although this is still suitable for amplification. In contrast, cervical smears and sputum samples contain very large numbers of cells. Also, exfoliative cytology specimens are usually rapidly fixed in an ethanol-based fixative which preserves the quality of the DNA, and produces very good yields of high-molecular-weight DNA (see *Figures 4 and 5*) which is an excellent template for PCR.

The ability to use archival cervical smears provides material for the investigation of cervical neoplasia, and also for epidemiological population studies. It also has important implications for forensic medicine, as smears are available from a large number of women, and may allow the genetic fingerprinting and subsequent matching of biological material found at crime scenes to archival hospital tissue from that individual. The only disadvantage of this technique is that it obviously destroys the cytology specimen.

4.4 Gross museum specimens

Interesting gross pathological specimens for museum display are fixed in 4% formaldehyde, containing potassium iodide and potassium acetate (K1 solution), then passed through absolute alcohol (K2 solution) before final storage in Perspex or glass containers in 25% glycerine, again containing potassium iodide and acetate at 5 g/litre (K3 solution). The potassium salts in these solutions help the specimens hold their colour.

Such specimens often include those from unusual or rare conditions, and it may be interesting to study their nucleic acid composition. Small blocks can be cut from the specimen without spoiling its appearance, and we have been able to extract DNA from these by 5-day proteinase K incubation. The tissue can either be crushed, minced or embedded in paraffin wax for sectioning prior to digestion.

David P. Jackson, Jeremy D. Hayden, and Phil Quirke

5. Extraction and amplification of RNA

DNA can be amplified from a single-stranded RNA template by combining a standard PCR protocol with an initial incubation with a retrovirus reverse-transcriptase enzyme (refs 10 and 11; see also Chapters 7, 10, and 12). However, the extraction of undegraded template RNA from tissues is particularly difficult because of its labile nature, and the presence of active ribonucleases in the tissue itself. Digestive enzymes including RNases are synthesized on ribosomes on the cytoplasmic surface of the endoplasmic reticulum, then passed through into the internal compartment before being packaged into secretory granules. This process effectively separates the synthetic pathways of the cytoplasm from the degradative function of these enzymes. However, disruption of the cell during an extraction procedure would allow RNases to act on nuclear and cytoplasmic RNA. RNases are also very active enzymes that do not require cofactors to function (11).

Standard RNA extraction protocols (2, 12), overcome this problem by lysing cells in guanidium thiocyanate (13) a chemical environment that results in the denaturation of RNases. The RNA must then be purified either by organic extraction and ethanol precipitation or sedimentation through caesium chloride. Because RNases are present in the environment, these procedures require the use of sterile plastic and glassware. Hence the isolation of pure, intact RNA is a complicated procedure. Despite this, the sensitivity and specificity should mean the reverse-transcriptase PCR (RT-PCR) is able to tolerate small quantities of poor quality, partly degraded RNA, and less complex extraction procedures should suffice.

We have found that a 5-day proteinase K incubation for DNA extraction from paraffin-embedded tissue also produces significant amounts of RNA and, as with DNA, of larger amounts than SDS incubation or boiling tissue. Although RNA is difficult to quantify, it can be demonstrated on gel electrophoresis of the extracted nucleic acid sample, before and after digestion with RNase (*Figure 6*). Also, such RNA can be produced without any special treatment of equipment or solutions. However, we have found that it is essential to add glycogen, as a carrying agent, during the ethanol/sodium acetate precipitation step. We have shown previously that RNA extracted by this method is suitable for PCR amplification and the detection of viral RNA sequences (14). Long-term storage of such RNA in solution is not recommended because it is likely to be broken down by RNases introduced during the extraction procedure and subsequent handling of the sample.

Protocol 3. Reverse transcriptase PCR

1. Combine in a microcentrifuge tube:
- 50 pmol of primer
- 10 mmol of each dNTP (BRL)   1 µl

Protocol 3. *Continued*

- Moloney murine leukaemia virus (MLV) superscript
- RT (BRL; 200 units/ml)   1 μl
- RNasin (Promega; 10 units/μl)   1 μl
- 10 × PCR buffer (Promega or see below)[a]   1 μl

  Make up to a final volume of 10 μl with dH_2O.

2. Mix by agitation and microfuge for 10 sec at 13 000 g
3. Place in thermal cycler (e.g. Grant Autogene) and incubate at 42 °C for 30 min.
4. Heat kill MLV-RT at 94 °C for 5 min.
5. Remove from thermal cycler, quick chill on ice at 4 °C.
6. Add:
   - 50 pmol of second primer
   - 10 × PCR buffer (Promega or see below)[a]   4 μl
   - deionized water   34.5 μl
   - *Taq* polymerase   0.5 μl
7. Overlay with liquid paraffin and thermocycle at 94 °C for 40 sec, 55 °C for 45 sec, and 72 °C for 1 min for 40 cycles. Complete cycling by 5 min at 72 °C to fully extend DNA. Cool to 4 °C and the products are ready for running on a 3% agarose gel.

[a] 10 × PCR buffer: 500 mM KCl, 100 mM Tris–Cl, pH8.8, 0.1% gelatin, 1% Triton X-100 and $MgCl_2$, usually at 1.5 mM.

---

6. Improving the sensitivity and specificity of PCR amplification

6.1 Nested PCR

Problems with PCR sensitivity do arise, especially when dealing with poor quality or low-copy-number nucleic acid template. Using nested PCR (nPCR) or nested RT-PCR (nRT-PCR) the sensitivity and specificity of both DNA and RNA amplification is considerably improved.

The process (15, 16) utilizes two consecutive PCRs, each usually involving 25 cycles of amplification. The first PCR contains an external pair of primers, while the second contains two nested primers which are internal to the first primer pair, or one of the first primers and a single nested primer. The larger fragment produced by the first reaction is used as template for the second PCR. The result is effectively 50 cycles of amplification of the fragment flanked by the nested primer pair (see *Figure 7*). In our hands, however, nPCR can be 1000 times more sensitive than 50 cycles of standard PCR. This is probably due to the more effective denaturation of the smaller fragments

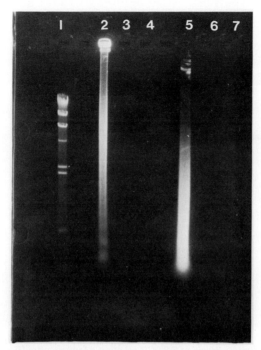

Figure 6. Nucleic acid extracted from sections of paraffin-embedded tonsil using a variety of extraction methods showing the superior yield of RNA from a 5-day proteinase K incubation. **Lane 1**: λ *Hind*III digest; **lane 2**: Nucleic acid extracted by proteinase K incubation, after RNase digestion; **lane 3**: Nucleic acid extracted by SDS incubation, after RNase digestion; **lane 4**: Nucleic acid extracted by boiling, after RNase digestion; **lane 5**: Nucleic acid extracted by proteinase K incubation, after DNase digestion; **lane 6**: Nucleic acid extracted by SDS incubation, after DNase digestion; **lane 7**: Nucleic acid extracted by boiling, after DNase digestion.

added to the second reaction from the first round of amplification (16).

Using nRT-PCR we have successfully gone on to perform single- and double-stranded sequencing of Measles virus RNA from archival paraffin embedded tissue (see *Figure 8*). Single-stranded sequencing was performed using asymmetric PCR (see Chapter 2). Otherwise the method used was a standard dideoxy sequencing method.

6.2 'Hot' nested PCR

When required, the sensitivity of nested PCR can be enhanced further using 'hot' nPCR. This technique elegantly combines the qualities of nPCR with the high resolution of polyacrylamide gel electrophoresis, resulting in detection to the order of 1 molecule in 10^6 cells. This facilitates the unequivocal detection of very rare sequences and is ideally suited to the study of persistent viral infections (15).

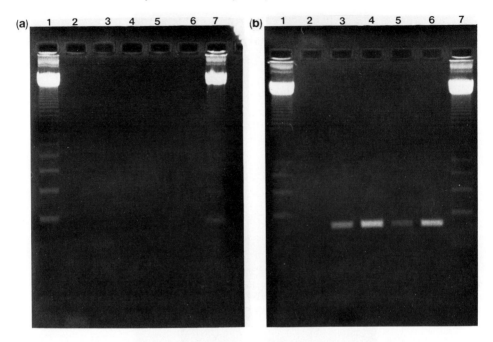

Figure 7. (a) Gel electrophoresis of RT-PCR reaction products after 25 cycles of amplification of a 245 bp measles virus RNA fragment from archival material showing no visible product. (b) Gel electroporesis of 194 bp products generated by a further 25 cycles of nested amplification using internal primers. **Lane 1**: 123 bp ladder standard; **Lane 2**: Negative reagent control (distilled water): **Lane 3**: Brain tissue from a characterized case of subacute sclerosing panencephalitis (SSPEn); **Lane 4**: Brain tissue from a suspected case of SSPEn; **Lane 5**: Spleen tissue actively infected with wild-type measles virus; **Lane 6**: RNA extracted from the measles, mumps, rubella attenuated live trivalent vaccine.

The technique is based on a nested PCR with the inclusion of an end-labelled primer in the second reaction. End-labelling can be conveniently performed by incubating one of the primers with polynucleotide kinase and $[\gamma\text{-}^{35}S]$ ATP (*Protocol 4*).

Protocol 4. Primer end-labelling

For 10 'hot' PCRs, using 50 pmol of primer in each PCR.

1. Mix and incubate the following at 37°C for 45 min:

- 500 pmol dried down internal primer
- 4 μl of T4 kinase (BRL)
- 3 μl of kinase buffer (BRL)
- 3 μl of ^{35}S-ATP
- 20 μl distilled water
- Final volume 30 μl

2. Inactivate kinase by heating to 70°C for 2 min.

3. Clean up primer when cool by running through a Biogel P6 (Bio-Rad) column.

4. Store at −20°C until required.

5. Add 3 μl of 'hot' primer into a PCR after adding the sample template.

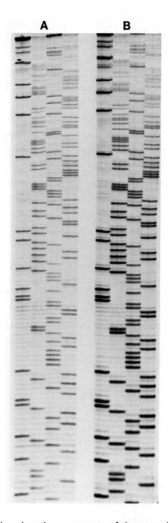

Figure 8. Sequencing gel showing the sequence of the measles virus nucleocapsid gene from a cDNA and from RNA extracted from a brain biopsy from a case of SSPEn which had been formalin fixed and paraffin-embedded. Sequence was determined from single-stranded DNA generated by asymmetric PCR from either a DNA-PCR (cDNA) or RT-PCR (SSPEn).

The resulting 'hot' nPCR products are then resolved on a high-resolution polyacrylamide denaturing gel. The detection of the fragment of the desired length can then be confirmed with precision.

6.3 Avoiding PCR contamination

The exquisite sensitivity of PCR often results in problems with reagent or sample contamination. This occurs when exogenous previously amplified sequences, positive control plasmids or DNA from the skin surface of the PCR operator contaminate the reaction mixture resulting in a false-positive signal. Procedures which aim to enhance the sensitivity of PCR may potentiate this problem. A number of routine precautions should be taken when performing any type of PCR to eliminate the unwanted transfer of DNA and the spectre of PCR contamination (17).

Ideally, experiments should be set up in a laminar flow hood dedicated solely for PCR use. This should be situated in a separate laboratory completely isolated from PCR product stores or plasmid clone preparation areas. Separate supplies of pipettors, pipette tips, Eppendorf tubes, and reagents should be kept specifically for PCR and not be used for any other preparations. Reagents and equipment such as PCR buffer, distilled water, pipette tips, and Eppendorfs should be autoclaved.

Oligonucleotide primers should be aliquoted once synthesised, preferably in a 'clean' laboratory. Due to the cost of replacement, special care should be taken with primers; the use of a separate box of pipette tips when aliquoting them is advisable. Disposable gloves should be worn at all times and changed frequently, especially when entering or leaving the PCR laboratory, handling concentrated DNA samples or when a splash occurs. Sample DNA or RNA should be added to the reaction last, and it is a good idea to change gloves in between each sample. The spread of contamination picked up by the PCR operator may also be avoided by the wearing of protective clothing which covers the head and face (18).One or more negative reagent controls containing distilled water instead of template should always be included when performing PCR so that carelessness in preparation can be highlighted. An extraction control is valuable when dealing with paraffin-embedded material. This comprises a negative control tissue block which is extracted alongside the test samples, thus controlling for contamination prior to setting up the PCR.

Positive control DNA plasmids should be diluted down considerably and added in a different laboratory. Only a very small volume would suffice to contaminate the PCR preparation area and very little is needed to produce a strong positive control signal.

Nested PCR can also be adopted as an anti-contamination strategy. In addition to its enhanced sensitivity and specificity this method avoids false-positive signals arising as a result of 'jumping PCR' (19). This is where partially overlapping fragments of DNA present in the reaction mixture may extend along each other to form an amplifiable product of the correct length giving a false-positive result.

6.4 Ultraviolet-mediated DNA crosslinking

With patient adherence to these good habits, most contamination problems should disappear. If, however, false-positives still arise then UV DNA crosslinking may provide the solution (20).

Before adding the sample template into a PCR mixture, the contents can be exposed to short-wave (254 and 300 nm together or just 254 nm) UV radiation (*Protocol 5*). Such an attack should sufficiently nick and crosslink any contaminating sequences rendering them unamplifiable (see *Figure 9*).

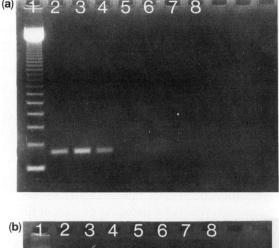

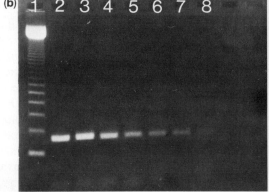

Figure 9. Effect of short-wave ultraviolet radiation exposure on the amplification of varying quantities of measles virus cDNA. Comparison of products produced with (a) and without exposure to 80 min of UV irradiation prior to amplification (b). **Lane 1**: 123 bp ladder standard; **lane 2**: 100 ng; **lane 3**: 50 ng; **lane 4**: 25 ng; **lane 5**: 2.5 ng; **lane 6**: 250 pg; **lane 7**: 25 pg; **lane 8**: 2.5 pg.

Protocol 5. UV DNA cross-linking to destroy exogenous template

1. Set up a PCR (as described in other chapters), but do not add sample DNA.
2. Transfer PCR tubes to a UV crosslinker (for example, Hoefer Scientific Instruments, model UVC 1000).
3. Set the desired duration of exposure[a] and follow all manufacturers' safety instructions for operating a short-wave UV source.
4. Expose the PCR tubes by opening lids.
5. Remove from crosslinker, add sample nucleic acid, seal with liquid paraffin, transfer to thermal cycler, and commence PCR.

[a] The length of exposure depends on the quantity of likely contamination which in turn depends on the presence of concentrated target DNA, e.g. plasmids, in the laboratory.

In our experience the disadvantage with this procedure is that the primers and *Taq* polymerase are adversely affected by long exposure to UV, resulting in a subsequent loss of sensitivity to the reaction. If primers can be ruled out as a source of contamination then they need not be exposed for long periods. Following 80 min of UV radiation exposure, up to 25 ng of plasmid DNA cannot be detected by 40 cycles of PCR amplification using our equipment. Shorter exposure times will deal with lesser amounts of contamination. Higher intensity UV sources will require less exposure.

Routine exposure of PCR buffer, dNTPs, and distilled water to short-wave UV radiation is recommended as an anti-contamination precaution. Exposure of *Taq* polymerase and primers for excessive periods of time is recommended only as a last resort. The sensitivity of the assay will then be compromised.

7. Summary

DNA for use as template material with PCR can be extracted, not only from fresh tissue and cytology specimens, but also from a wide range of archival preparations, including fixed paraffin-embedded tissue, H&E-stained sections, gross museum specimens, and a variety of archival cytology preparations. This can be performed cheaply and conveniently, using standard proteinase K incubation methods, which can produce DNA from very small quantities of tissue. Alternatively, where larger amounts of tissue are available DNA can be extracted by simply boiling the tissue in distilled water. The use of an alcohol-based fixative greatly improves the yield and quality of DNA that can be extracted. RNA can also be extracted from archival material by incubation with proteinase K.

The sensitivity of PCR for the amplification of DNA from such specimens can be increased by using nested PCR primers and end-labelled primers.

When using PCR a number of precautions must be used to prevent false-positive results due to contamination. UV crosslinking of the PCR mix may be useful if contamination is a consistent problem.

The ability of PCR to amplify poor quality genetic material from previously uninvestigated archival tissue and cytology preparations makes it a very valuable technique.

Acknowledgements

We would like to thank Mrs J. Fearnley for typing this chapter. D. P. Jackson and J. D. Hayden were supported by Jean Shanks Fellowships. This work has been supported by grants from the Yorkshire Cancer Research Campaign, British Society of Clinical Cytology, and the Yorkshire Region Locally Organised Research Scheme. We are grateful to J. A. Hoyle for kindly providing *Figure 8*.

References

1. Higuchi, R., von Beroldingen, C., Sensabaugh, G. F., and Erlich, H. A. (1988). *Nature*, **332**, 843.
2. Maniatis, T., Fritsch, E. F., and Sambrook, J. (1982). *Molecular cloning: a laboratory handbook*. Cold Spring Harbor Laboratory Press, Cold Spring Harbor, NY.
3. Lench, N., Stanier, P., and Williamson, R. (1988). *Lancet*, **i**, 1356.
4. Jackson, D. P., Lewis, F. A., Taylor, G. R., Boylston, A. W., and Quirke, P. (1990). *Journal of Clinical Pathology*, **43**, 499.
5. Warford, A., Pringle, J. H., Hay, J., Henderson, S. D., and Lauder, I. (1988). *Journal of Pathology*, **154**, 313.
6. Goelz, S. E., Hamilton, S. R., and Vogelstein, B. (1985). *Biochemical and Biophysical Research Communications*, **130**, 118.
7. Lo, Y-M. D., Mehal, W. Z., and Fleming, K. A. (1989). *Journal of Clinical Pathology*, **42**, 840.
8. Pääbo, S. (1990). *PCR protocols: A guide to methods and applications* (ed. M. A. Innis, D. H. Gelfand, J. J. Sninsky, and T. J. White), p. 159. Academic Press, San Diego, California.
9. Jackson, D. P., Bell, S., Payne, J., Lewis, F. A., Sutton, J., Taylor, G. R., and Quirke, P. (1989). *Nucleic Acids Research*, **17**, 10134.
10. Hart, C., Spira, T., Moore, J., Sninsky, J., Schochetman, G., Lifson, A., Galphin, J., and Ou C-Y. (1988). *Lancet*, **ii**, 596.
11. Lynas, C., Cook, S. D., Laycock, K. A., Bradfield, J. W. B., and Maitland, N.J. (1989). *Journal of Pathology*, **157**, 285.
12. Kingston, R. E. (1987). In *Current protocols in molecular biology* (ed. F. M. Ausubel, R. Brent, R. E. Kingston *et al.*), p. 4.2.3. Green Publishing Associates and Wiley Interscience.
13. Chirgwin, J. M., Przybyla, A. E., MacDonald, R. J., and Rugger, W. J. (1979). *Biochemistry*, **18**, 5294.

14. Jackson, D. P., Quirke, P., Lewis, F., Boylston, A. W., Sloan, J. M., Robertson, D., and Taylor, G. R. (1989). *Lancet*, **i,** 1391.
15. Simmonds, P., Balfe, P., Peutherer, J. F., Ludlam, C. A., Bishop, J. O., and Brown, A. J. L. (1990). *Journal of Virology*, **64,** 864.
16. Porter-Jordan, K., Rosenberg, E. I., Keiser, J. F., Gross, J. D., Ross, A. M., Nasim, S., and Garrett, C. T. (1990). *Journal of Medical Virology*, **30,** 85.
17. Kwok, S. and Higuchi, R. (1988). *Nature*, **339,** 237.
18. Kitchin, P. A., Szotyori, Z., Fromholc, C., and Almond, N. (1990). *Nature*, **344,** 201.
19. Porter-Jordan, K. and Garrett C. T. (1990). *Lancet*, **335,** 1220.
20. Sarkar, G. and Sommer, S. S. (1990). *Nature*, **345,** 27.

4

Analysis of genomic sequence variation using amplification and mismatch detection (AMD) and direct sequencing

ROLAND G. ROBERTS, A. JANE MONTANDON,
PETER M. GREEN and DAVID R. BENTLEY

1. Introduction

Genomic DNA sequence variation between individuals is the basis of both heritable phenotypic traits (including lethal diseases) and the neutral poly-morphisms used for linkage analysis (1). The analysis of new disease mutations has until recently been performed by cloning and sequencing (e.g. see ref. 2), while the identification of novel restriction fragment length polymorphisms (RFLPs) for a given DNA probe has usually been effected by Southern analysis of DNAs digested with a selection of restriction enzymes (1, 3). The advent of the polymerase chain reaction (4) (PCR), direct sequencing (5), and mismatch detection [using chemical modification (6, 7), RNase A cleav-age (8), or denaturing gradient gel electrophoresis (9)] has improved both the speed and efficiency of detecting all types of sequence variation. This chapter describes the use of PCR in combination with chemical mismatch detection and sequencing to detect sequence variation, and discusses their application to the study of both mutations and polymorphisms.

2. Amplification and mismatch detection (AMD) analysis

AMD analysis (7) (see *Figure 1*) is a combination of PCR (4) and the chemical mismatch method (6). The region of interest (an exon of a gene or a segment of an anonymous probe, for example) is amplified from the genomic DNA of the individuals to be examined (*Protocol 1*). The PCR product from a refer-ence sample (e.g. for detection of a mutation, DNA from a normal individual might be used) is then radioactively end-labelled (*Protocol 2*) and mixed with

(a)

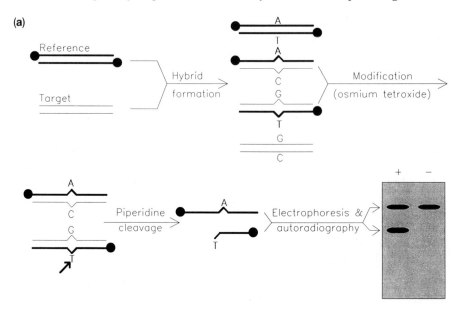

(b)

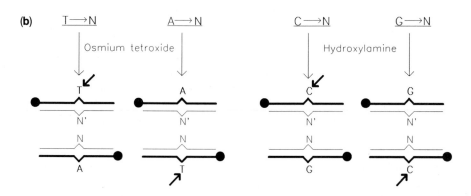

Figure 1. (a) Schematic diagram of AMD analysis. In the example, a mismatch formed by a target sample which has a A→G transition with respect to the labelled (filled circles) reference sample is detected using osmium tetroxide modification. The schematic auto-radiograph shows bands expected when the target sample is different from (+) or the same as (−) the reference sample. (b) AMD analysis can detect any substitution mutation. Any deletion or insertion presents one or more mismatched pyrimidines in at least one strand which disrupt the neighbouring (formally paired) bases sufficiently to allow chemical modification to occur (21).

an excess of the PCR products from each of the other individuals. The mixtures are denatured and then allowed to renature to form hybrid duplexes (*Protocol 3*). Any variation between the 'probe' sequence and the other strand in the duplex will disrupt base-pairing; this site is sensitive to specific chemical modification by either hydroxylamine (10) (mismatched cytosine residues) or osmium tetroxide (11) (mismatched thymidine residues). The DNA strand can then be cleaved at the site of modification with piperidine and analysed by denaturing polyacrylamide gel electrophoresis and auto-radiography (*Protocol 4*). We generally use probes labelled at both ends, thus analysing both strands simultaneously. Alternatively, the use of single end-labelled probes (see below) allows unequivocal localization of the sequence variation. The distance between the point(s) of sequence variation and the end-label will then be represented by the size of the cleavage product(s). An example of AMD analysis is shown in *Figure 2*.

2.1 PCR amplification

The limiting factor on PCR product size when used for AMD analysis is the resolution of the larger products of the mismatch reaction on the acrylamide gel, where the ability to localize the position of the mismatched site is lost. Other gel systems might be used to raise this limit.

Oligonucleotides should be chosen taking precautions to avoid short re-peated sequences, A/T-rich stretches, and regions of complementarity be-tween a primer and either itself or its PCR partner. We use primers with lengths of 20–25 bp. Rare-cutter restriction sites can be incorporated into the 5'-ends of the primers if the probe is to be labelled at the 3'-end in a fill-in reaction (*Protocol 2*; ref. 1) (this will allow specific labelling of one end of the AMD probe): a *Sal*I site on one primer and a *Cla*I site on the other, for instance. To ensure efficient restriction, a further 2–3 bp are included 5' of the recognition site.

The method for PCR and subsequent purification of products using Gene-clean (Bio 101 Inc, Stratech Scientific) is described in *Protocol 1*. The purifica-tion results in removal of both PCR primers and unincorporated nucleotides, which would otherwise interfere with labelling or sequencing and are also likely to affect the mismatch signal by serving as additional targets for chemi-cal modification.

Protocol 1. PCR and product purification

1. Make up the reaction pre-mix as follows (these quantities for one sample):
- 10 × PCR buffer (*Table 1*)   5 µl
- 5 mM dNTPs   5 µl
- 100 ng/µl Primer 1   5 µl
- 100 ng/µl Primer 2   5 µl

Analysis of sequence variation by AMD and sequencing

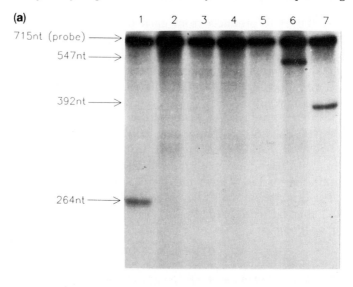

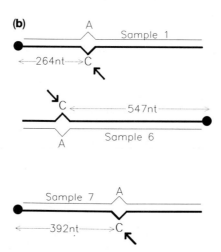

Figure 2. (a) AMD analysis of seven haemophilia B patients using a PCR reaction lying in exon h(7). Three of the lanes show mismatch bands due to differences between the targets and the (normal) reference sample. 3′-end labelling and hydroxylamine were used. (b) Schematic diagram showing how the mutations give rise to their respective AMD products. The mutations are London 13, Malmö 7, and London 5 respectively (data from ref. 7)

Protocol 1. *Continued*

- 5 mg/ml bovine serum albumin (BSA)   1.7 μl
- 5% v/v β-Mercaptoethanol   0.7 μl
- 5U/μl AmpliTaq *Taq* polymerase (Cetus)   0.6 μl
- $T_{0.1}E$ 17 μl

2. Add 40 μl of this mixture to 10 μl of each genomic DNA (approx. 10 ng/μl in $T_{0.1}E$) in a 0.5 ml Eppendorf tube. Cover sample with 50 μl of light paraffin, mix well, and place in a programmable heating block.

3. Heat the reaction at 94°C for 5 min, then subject to 30 cycles of:

   - 93°C for 1 min
   - 55–65°C (depending on primers) for 1 min
   - 72°C for 2–7 min (depending on length of product) followed by 5 min at 72°C.

4. Analyse 5–10 μl of the product by electrophoresis in a 5% polyacrylamide (or 4% NuSieve agarose) minigel containing ethidium bromide.

5. Purify the remaining product using Geneclean (Bio 101 Inc.),[a] following the manufacturer's instructions and eluting in 10–20 μl of $T_{0.1}E$. Estimate the concentration of the product by minigel electrophoresis.

---

[a] Other methods, such as gel purification (12) or Sepharose CL-6B (linker) columns (Boehringer-Mannheim) may also be considered.

2.2 Labelling of probe sample

The purified PCR product which is chosen as the probe must now be end-labelled (*Protocol 2*)—approximately 5 ng of probe DNA per target sample per modification reaction are needed. Radioactive labelling of the probe is done either using T4 polynucleotide kinase (5′-end-labelling) or by restriction followed by fill-in labelling using the Klenow fragment of DNA polymerase I (3′-end). 5′-end labelling has the advantage that it can be carried out on any amplified sample. Selective labelling of one strand is not possible. 3′-end labelling can be used to selectively label each strand of the probe separately, but this requires that the appropriate restriction sites are present in the oligonucleotide primers. A further disadvantage is that it is not possible to confirm efficient digestion of the probe before labelling. After labelling by either method, an aliquot of the probe may be restricted and electrophoresed in a 6% denaturing polyacrylamide gel in order to allow assessment of the degree of labelling at each end. If the labelling reaction gives a smear below the cognate band, or yields lower-molecular-weight contaminants, it may be advisable to gel-purify the probe before using it for AMD analysis (12).

Protocol 2. Labelling of probe

Label by either:

A. *3′-end labelling*

1. Cleave the terminal restriction site(s) with the appropriate enzyme(s).

2. Mix:

• Probe DNA	x μl
• [α-^{32}P]-dNTP(s) (3000 Ci/mmol; 10 Ci/ml)	1 μl
• TM	1 μl
• 0.1 M dithiothreitol (DTT)	1 μl
• 5 units/μl Klenow fragment (Boehringer-Mannheim)	0.5 μl
• $T_{0.1}E$	(6.5 − x) μl

3. Incubate at r.t. for 15 min. Place on ice or store at −2°C.

or:

B. *5′-end labelling*

1. Mix:

• Probe DNA	x μl
• [γ-^{32}P]-ATP (3000 Ci/mmol; 10 Ci/ml)	0.5 μl
• 10 × kinase buffer (freshly made: *Table 1*)	1 μl
• 10 units/μl T4 polynucleotide kinase	0.5 μl
• $T_{0.1}E$	(8 − x) μl

2. Incubate at 37°C for 30 min. Place on ice or store at −20°C.

2.3 Preparation of hybrids

To minimize the formation of probe homoduplexes, hybridizations between probe and target are set up in which the target DNA is present in 10- to 20-fold molar excess over probe DNA. Quantities are described in *Protocol 3* for analysis of eight samples with both osmium tetroxide and hydroxylamine.

Protocol 3. Hybrid preparation

1. Make a pre-mix containing 18 μl of 10 × hybrid buffer (*Table 1*) and the appropriate quantity of labelled probe (approx. 90 ng) from *Protocol 2*, making up to a total of 162 μl with $T_{0.1}E$.

2. Add 18 μl of pre-mix to 2 μl of each target sample (i.e. approx. 100–200 ng) in 0.5 ml Eppendorf tubes. Top with 50 μl of paraffin. Place in boiling water bath for 5 min.

3. Immediately transfer tubes to 65°C water bath (do not allow to cool below 65°C). Incubate for 5–16 h to allow hybrid formation to occur.

4. Transfer aqueous phase to siliconized 1.5 ml Eppendorfs. Add 3 μl of 20 mg/ml mussel glycogen (Boehringer-Mannheim) and 750 μl of stop/precipitation mixture (*Table 1*). Mix well, chill on dry-ice for 10 min.

5. Spin at ~14 000 g in a microcentrifuge for 10 min. Discard the supernatant, rinse the pellet with 70% v/v ethanol, and resuspend in 14 μl of $T_{0.1}E$. The samples may be stored at −20°C overnight or used immediately as substrates for chemical modification by osmium tetroxide and/or hydroxylamine (*Protocol 4*).

Table 1. Composition of solutions

● 10 × PCR buffer		670 mM	Tris–HCl pH 8.8
		166 mM	Ammonium sulphate
		67 mM	Magnesium chloride
● $T_{0.1}E$		10 mM	Tris–HCl pH 7.4
		0.1 mM	EDTA
● Formamide dyes		95%	Deionized formamide
		10 mM	EDTA
		10 mg/ml	Xylene cyanol
		10 mg/ml	Bromophenol Blue
● TM		100 mM	Tris–HCl pH 8.3
		50 mM	Magnesium chloride
● 10 × Kinase buffer		500 mM	Tris–HCl pH 7.6
		100 mM	DTT
		100 mM	Magnesium chloride
● 10 × Hybrid buffer		3 M	NaCl
		1 M	Tris–HCl pH 8.0
● Stop/precipitation mixture		63 mM	Sodium acetate
		20 μM	EDTA
		80%	Ethanol
● Chase solution		250 μM	dGTP, dATP, dTTP, dCTP
		10%	DMSO
● Termination mixes	G:	80 μM	dGTP, dTTP, dCTP
		8 μM	ddGTP
	A:	80 μM	dGTP, dTTP, dCTP
		0.8 μM	ddATP
	T:	80 μM	dGTP, dTTP, dCTP
		8 μM	ddTTP
	C:	80 μM	dGTP, dTTP, dCTP
		8 μM	ddCTP

2.4 Mismatch analysis

It is convenient to perform operations outlined in steps 2A and 2B of *Protocol 4* concurrently, as they differ only in their initial stages. Note that many of the chemicals used in this and the subsequent section are highly toxic, and should be handled and disposed of according to manufacturers' instructions.

Protocol 4. Chemical modification, cleavage, and analysis

1. Split each hybrid sample (from *Protocol 3*) into 2 × 7 μl aliquots in siliconized 1.5 ml Eppendorfs (one for each chemical modification procedure).

2A. Osmium tetroxide only:

  (a) Make a fresh solution of osmium tetroxide and pyridine by mixing:

      • Pyridine   6.75 μl

      • 4% w/v osmium tetroxide[a]   1.44 μl

      • $T_{0.1}E$ 154 μl

  (b) Add 18 μl of this solution to each of the 7 μl samples of hybrid from step 1, mix well and incubate at 37°C for 2 h (final concentrations of reagents are 0.025% OsO_4 and 3% pyridine).

2B. Hydroxylamine only:

  (a) Make a 4 M solution of hydroxylamine hydrocholoride adjusted to pH 6 with diethylamine:

      • Deionized water   138 μl

      • Diethylamine   42 μl

      • Hydroxylamine hydrochloride   39 mg

  (b) Add 20 μl of this solution to each of the 7 μl samples of hybrid from step 1. Mix well and incubate at 37°C for 2 h (giving a final hydroxylamine concentration of 2.3 M)

3. Add 750 μl of stop/precipitation mixture to each tube from steps 2A and 2B. Mix well, chill on dry-ice for 10 min. Spin at ~14 000 g in a microcentrifuge for 10 min. Discard the supernatant and rinse the pellet with 70% ethanol. Centrifuge briefly (5 sec) and remove all remaining liquid from the tubes.

4. Make a fresh 1 M solution of piperidine[b] in deionized water (90 μl piperidine + 810 μl water, for this sample number). Mix on ice and keep the tubes open for the shortest possible time. Add 50 μl of this solution to each sample tube and vortex to resuspend the pellet. Incubate at 90°C for 30 min[b] in, for example, a paraffin-filled dry-block). Place a weight on top of the tubes to ensure they remain sealed.

5. Add 750 μl of stop/precipitation mixture to each tube. Mix well, chill on

dry-ice for 10 min. Spin at ~14000 g in a microcentrifuge for 10 min. Discard the supernatant and rinse the pellet with 70% ethanol.

6. Resuspend the pellet in 7 µl of a 1/7 dilution of stock formamide dyes, boil, snap-chill, and load on to a denaturing 4–6% polyacrylamide gel. Electrophorese and autoradiograph.

[a] Use for up to two months. Aldrich supply osmium tetroxide as a 4% solution (Cat. No. 25175–5).
[b] Piperidine is a controlled substance in many countries, but according to Shi *et al.* (13), it can be substituted with no adverse effects by pyrrolidine. The 90°C incubation should be reduced to 15 min.

---

3. Direct sequencing

The method for sequencing PCR products described below uses the double-stranded product directly as a template for sequence analsis of either strand. Primers used for the preceding PCR or internal primers can be used to initiate the sequencing reaction. This approach is simpler than the two published methods which involve the generation of a single-stranded template. In asymmetric PCR, for example, an excess of one PCR primer is used to drive predominant synthesis of one strand (14). The 'genomic amplification with transcript sequencing' method (GAWTS, ref. 15) uses a bacteriophage promoter in one primer to transcribe a template for reverse transcriptase sequencing. For either of these two methods, it is necessary to prepare template separately for analysis of each strand.

This protocol (5) takes the form of a traditional dideoxy chain termination reaction (16), modified by adjustment of primer concentration and annealing conditions to favour primer/template annealing over template–template annealing. Dimethyl sulphoxide (DMSO) is also used in the buffer (17). The method can be used for sequencing both PCR products and other linear double- or single-stranded DNA templates.

Quantities are presented in *Protocol 5* for the sequencing of eight template samples. The reaction is performed in a round-bottomed 96-well microtitre plate for ease of handling (ensure that the particular make of plate used can withstand boiling without deformation, e.g. Nunc 2–62170).

Protocol 5. Direct sequencing

1. Make primer pre-mix containing 5 µl of 100 ng/µl sequencing primer, 18 µl of 5 × Sequenase buffer (*Table 1*), 25 µl $T_{0.1}E$, and 6 µl of DMSO. Add 6 µl of this to 1 µl of Genecleaned PCR product (see *Protocol 1*; concentration usually 50–100 ng/µl) in 0.5 ml Eppendorf tubes.

2. Make enzyme pre-mix containing:
   - α-^{35}S-dATP (600 Ci/mmol; 10 Ci/ml)   4 µl

Protocol 5. *Continued*

- 100 mM DTT   8 μl
- DMSO   3 μl
- 12.5 units/μl Sequenase version 2.0 (USB)   1 μl
- $T_{0.1}E$ 19 μl

3. Dispense 2 μl of each of the 4 'termination mixes' per template into separate wells of a microtitre plate.[a,b]

4. Boil the mixtures of primer and templates from *step 1* for 3 min and snap-chill in an ice–water bath.

5. Add 4 μl of enzyme pre-mix to each tube of snap-chilled mixture. Mix well and distribute 2 μl of each mixture into each of the 4 termination-mix-containing microtitre wells,[b] taking care not to touch the drop of termination mix. Spin or tap the plate to mix the reaction components, and float it on the surface of a 37°C water bath for 5 min.

6. Add 2 μl of 'chase solution' (see *Table 1*) to each well.[b] Mix and incubate again at 37°C for 5 min.

7. Add 2 μl of formamide dyes to each well.[b] Mix. At this point the sequencing reactions may be stored overnight at −20°C.

8. Cover the wells of the microtitre plate with Saran wrap and float on a boiling water bath for 3 min. Snap-chill by plunging on to an ice–water bath. Load on to a denaturing 6% polyacrylamide gel.[c] Electrophorese, then fix gel by immersing in 10% methanol/10% acetic acid for 15 min. Transfer gel on to Whatman 3MM paper, dry at 80°C in a vacuum gel drier and autoradiograph.

[a] To prevent splashes when floating the plate in steps 5, 6, and 8, run a rim of insulation tape around the outside edge.

[b] A Hamilton syringe fitted with a repeating dispenser (PB600 plus tip adaptor, V. A. Howe) may be used for convenience.

[c] If a region of known sequence is being scanned for variations, it is suggested that when a large number of samples have been sequenced using the same primer, changes are easier to spot when all G-tracks, A-tracks, etc., are run adjacently than when the G, A, T, and C tracks from each sample are run together (see *Figure 3b*).

Note that the presence of primer dimers can interfere with sequencing of regions immediately adjacent to the primer (about the first 20 bp). If information on this region is required, then the template should be gel-purified.

4. Applications

4.1 Characterization of disease mutations

The availability of rapid methods to detect all types of sequence variation permits the development of a new strategy of direct diagnosis in inherited

(a)

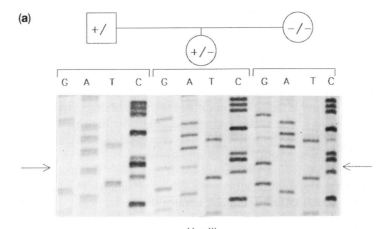

MaeIII
841Q+: ATGCAGTCACCATACAGCCC
841Q−: ATGCAGTCGCCATACAGCCC

(b)

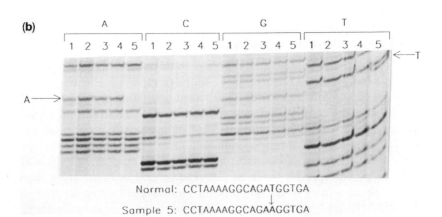

Normal: CCTAAAAGGCAGATGGTGA
Sample 5: CCTAAAAGGCAGAAGGTGA

Figure 3. (a) Direct sequencing of a region of the anonymous probe pERT84 shown by AMD analysis to vary between individuals (a 375 nucleotide band was generated in two out of the five samples used in the experiment). A clear G→A transition, which would result in hydroxylamine modification of the opposite strand during AMD analysis, is observed. This creates a *Mae*II site, which was later exploited (22). **(b)** Direct sequencing of a region of the Factor IX gene amplified from five haemophilia B patients, showing how a mutation is more readily identified when like tracks are run adjacently (5). Here sample 5 has an A→T transversion with respect to the normal sequence.

diseases in which the gene is known. Haemophilias A and B are examples of diseases which have a high frequency of new mutations which may occur at any location in the essential region of the gene. Previously, an indirect approach has been used for carrier and prenatal diagnosis at the DNA level, based on following the segregation of linked or intragenic RFLPs in each family. This approach is applicable to only approximately 60% of all cases. The remainder are either uninformative for the available markers, or they are isolated cases in which at best only exclusion of the high-risk chromosome is possible by RFLP analysis (18, 19). The use of linked markers for diagnosis also suffers the risk of misdiagnosis due to crossover between the marker and the mutation. A further disadvantage is the need for pedigree data and samples from key family members. Direct diagnosis by detection of the mutation, by contrast, is potentially applicable to every case, and suffers none of the disadvantages listed above.

Direct diagnosis has been used to screen the essential regions of the Factor IX gene of haemophilia B patients for mutations (5, 7). Analysis of DNA from relatives or from chorionic villus sampling has been used for carrier and prenatal diagnosis. Characterization of factor IX mutations in the population will also make it possible to construct a database of mutations in the population which can be used for more effective genetic counselling in the future.

4.2 Identification of novel polymorphisms

Previous approaches to detect new polymorphisms with a given probe have used Southern blotting. For example, Aldridge *et al.* (3) digested genomic DNA from three individuals with 23 different restriction enzymes. From the fifteen probes tested, however, only five detected polymorphisms. The main disadvantage with such a strategy is that only a small proportion of the available sequence variations is sampled: first, because only 10% of the genome lies within restriction sites, and second because there exist a large number of polymorphisms, such as CA repeats (20) or other short tandem repeats (21), of a form which could never be detected by this method.

Using AMD analysis, however, all possible sequence variations can be detected. For example, in order to detect polymorphisms which could be used as genetic markers, we analysed 1.4 kb and 1.8 kb regions at the 5′(22) and 3′(21) ends respectively of the dystrophin gene. Analysis of the 3′-end resulted in the detection in a single experiment of bands representing 42 distinct cleavage events in the reference DNA sample.

5. Discussion

5.1 Alternative mismatch detection methods

In addition to chemical modification and cleavage, there are several alternative published methods for detecting the presence of mismatched regions in heteroduplex molecules (see also Chapter 5).

5.1.1 RNaseA cleavage

In this method (8) a labelled RNA probe is annealed to the cognate DNA of interest and the probe is cleaved by the enzyme RNase A at mismatched pyrimidines. The method will localize the variable site in up to 70% of all cases. The efficiency of RNase A cleavage at different mismatches is variable, however, reaching only 5% completion at some U residues. Notwithstanding, it remains the only mismatch detection method to have been used successfully to detect point mutations in unamplified genomic DNA.

5.1.2 Denaturing gradient gel electrophoresis (DGGE)

The change in melting characteristics caused by a mismatch is exploited by monitoring the mobility shift of heteroduplexes in a denaturing gradient (9) (in 95% of cases). The method requires the presence of a 'GC-clamp' of up to 40 bp on at least one of the PCR primers and the use of specialized electrophoresis equipment. It fails to localize the variable site; the only information obtained is whether there is at least one site of sequence variation in the region or not.

5.1.3 Carbodiimide modification

Carbodiimide modification of mismatched G and T residues can induce electrophoretic retardation of DNA molecules up to 1.5 kb (23), but this suffers from the same disadvantage as DGGE since no information is obtained on the position of the variant site. Furthermore, use of this method has only been reported on plasmid DNA.

5.2 Designing PCR reactions for routine genotyping

After characterization of a polymorphism identified and AMD screening, it is preferable to replace the AMD assay with a rapid PCR reaction which is specific for the polymorphism under study.

If the polymorphism lies within a restriction enzyme site (22, 24) (*Figure 4a*), then a PCR reaction spanning the site can be designed. Amplification followed by restriction of the product will then yield a rapid assay for genotype at this locus.

If the polymorphism consists of a length change of 3 bp or more, this can be detected by designing a short reaction across the site of variation and resolving the products on a 12% polyacrylamide minigel (21, 25) (*Figure 4B*). In order to resolve length variations of 1 or 2 bp, radioactive labelling and vertical electrophoresis will be necessary (20).

If, however, neither of these approaches is possible, reactions using the principles of competitive oligonucleotide priming (COP; ref. 26) or amplification refractory mutation system (ARMS; ref. 27) may be used.

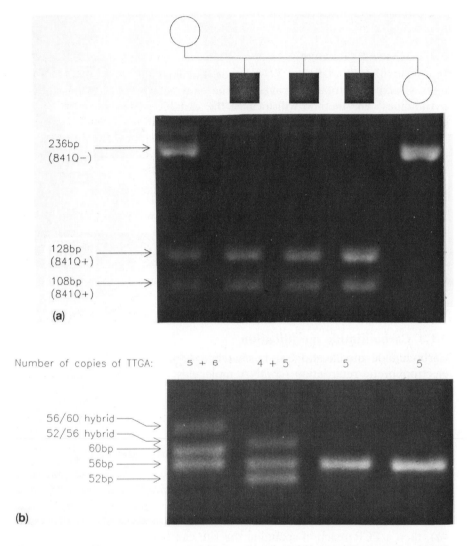

Figure 4. (a) A PCR reaction was designed to flank a polymorphic *Mae*III site in pERT84 identified by AMD analysis and direct sequencing (22) (see *Figure 3a*). Amplification of DNA from this family followed by *Mae*III digestion yielded a restriction pattern with Mendelian inheritance. (b) A PCR reaction spanning part of the 3′ untranslated region of the dystrophin gene allows analysis of the number of copies of a 4 bp repeated element by 12% PAGE. Although only 5- and 6-copy sequences were found in the initial AMD analysis (21), sample 2 has a rare 4-copy form. Note the hybrid band resulting from heterodimer formation in heterozygous individuals, which can itself be exploited as an analytical tool.

R. G. Roberts, A. J. Montandon, P. M. Green, and D. R. Bentley

Acknowledgements

This work has benefited from many helpful discussions with Professors Francesco Giannelli and Martin Bobrow. Some data presented derive from diagnostic work performed by Michael Yau and Stephen Abbs. Support for this work was received from the Medical Research Council, the Muscular Dystrophy Group of Great Britain and Northern Ireland, Action Research for the Crippled Child, the Spastics Society, and the Generation Trust.

References

1. Botstein, D., White, R. L., Skolnick, M., and Davis, W. D. (1980). *American Journal of Human Genetics*, **32**, 314.
2. Tsang, T. C., Bentley, D. R., Mibashan, R. S., and Giannelli, F. (1988). *EMBO Journal*, **7**, 3009.
3. Aldridge, J., Kunkel, L., Bruns, G., Tantravahi, U., Lalande, M., Brewster, T., Moreau, E., Wilson, M., Bromley, W., Roderick, T., and Latt, S. A. (1984). *American Journal of Human Genetics*, **36**, 546.
4. Saiki, R. K., Gelfand, D. H., Stoffel, S., Sharf, S. J., Higuchi, R., Horn, G. T., Mullis, K. B., and Erlich, H. A. (1988). *Science*, **239**, 487.
5. Green, P. M., Bentley, D. R., Mibashan, R. S., Nilsson, I. M., and Giannelli, F. (1989). *EMBO Journal*, **8**, 1067.
6. Cotton, R. G. H., Rodrigues, N. R., and Campbell, R. D. (1988). *Proceedings of the National Academy of Sciences USA*, **85**, 4397.
7. Montandon, A. J., Green, P. M., Giannelli, F., and Bentley, D. R. (1989). *Nucleic Acids Research*, **17**, 3347.
8. Myers, R. M., Larin, Z., and Maniatis, T. (1985). *Science*, **230**, 1242.
9. Sheffield, V. C., Cox, D. R., Lerman, L. S., and Myers, R. M. (1989). *Proceedings of the National Academy of Sciences, USA*, **86**, 232.
10. Johnston, B. H., and Rich, A. (1985). *Cell*, **42**, 713; Rubin, C. M., and Schmid, C. W. (1980). *Nucleic Acids Research*, **8**, 4613.
11. Friedmann, T., and Brown D. M. (1978). *Nucleic Acids Research*, **5**, 615.
12. Maniatis, T., Fritsch, E. F., and Sambrook, J. (ed.) (1982). *Molecular Cloning, A Laboratory Manual*. Cold Spring Harbor Laboratory Press, Cold Spring Harbor, NY.
13. Shi, Y., and Tyler, B. M. (1989). *Nucleic Acids Research*, **17**, 3317.
14. Gyllensten, U. B. and Erlich, H. A. (1988). *Proceedings of the National Academy of Sciences USA*, **85**, 7652.
15. Stoflet, E. S., Koeberl, D. D., Sarkar, G., and Sommer, S. S. (1988). *Science*, **239**, 491.
16. Sanger, F., Nicklen, S., and Coulson, A. R. (1977). *Proceedings of the National Academy of Sciences, USA*, **74**, 5463.
17. Winship, P. R. (1989). *Nucleic Acids Research*, **17**, 1266.
18. Giannelli, F. (1987). In *Protides of the biological fluids* (ed. H. Peeters), Vol. 35, pp. 29–32. Pergamon Press, Oxford.
19. Cole, C. G., Walker, A., Coyne, A., Johnson, L., Hart, K. A., Hodgson, S., Sheridan, R., and Bobrow, M. (1988). *Lancet*, **i**, 262.

20. Weber, J. L., and May, P. E. (1989). *American Journal of Human Genetics*, **44,** 388.
21. Roberts, R. G., Montandon, A. J., Bobrow, M., and Bentley, D. R. (1989). *Nucleic Acids Research*, **17,** 5961.
22. Roberts, R. G., Bobrow, M., and Bentley, D. R. (1990). *Nucleic Acids Research*, **18,** 1315.
23. Novack, D. F., Casna, N. J., Fischer, S. G., and Ford, J. P. (1986). *Proceedings of the National Academy of Sciences, USA*, **83,** 586.
24. Roberts, R. G., Cole, C. G., Hart, K. A., Bobrow, M., and Bentley, D. R. (1989). *Nucleic Acids Research*, **17,** 811.
25. Mathew, C. G., Roberts, R. G., Harris, A., Bentley, D. R., and Bobrow, M. (1989). *Lancet*, **ii,** 1346.
26. Gibbs, R. A., Nguyen, P.-N., and Caskey, C. T. (1989). *Nucleic Acids Research*, **17,** 2437.
27. Newton, C. R., Graham, A., Heptinstall, L. E., Powell, S. J., Summers, C., Kalsheker, N., Smith, J. C., and Markham, A. F. (1989). *Nucleic Acids Research*, **17,** 2503.
28. Cotton, R. G. H. and Campbell, R. D. (1989). *Nucleic Acids Research*, **17,** 4223.

Detection of deletions and point mutations

BELINDA J. F. ROSSITER, MARKUS GROMPE, and
C. THOMAS CASKEY

1. Introduction

Many human disorders result from the disruption of a single gene by mutation. In some genetic diseases, such as sickle cell anaemia, the same mutation
is found in many individuals and can be detected by relatively simple screening methods. In other genetic disorders, such as those resulting from defective
X-linked genes, mutations are heterogeneous and therefore require different,
scanning, approaches for their detection.

Until fairly recently it was only possible to search for mutations within
specific genes using the techniques of Southern and Northern analysis, and
sequencing of cDNA recombinants derived from mRNA. Such techniques
are time-consuming and costly and are not able to detect all mutations. The
development of the polymerase chain reaction (PCR) (1) has revolutionized
the area of mutation detection, not only accelerating the process, but also
making possible approaches which were previously impractical.

In this chapter we will describe methods for detecting almost all mutations
within the human hypoxanthine guanine phosphoribosyltransferase (HPRT)
gene, which when disrupted results in the human disorders gouty arthritis and
Lesch–Nyhan syndrome (2). Individual exons of the gene are simultaneously
amplified using the PCR (a 'multiplex' reaction) and visualized by agarose gel
electrophoresis. Deletions affecting all or part of the gene result in missing
bands within the multiplex PCR, and a more accurate determination of
deletion end-points than is possible with Southern analysis. Since not all
mutations within the HPRT gene are deletions, the products of the PCR
amplification are then sequenced directly (i.e. without further subcloning) to
determine point mutations and other small rearrangements (3). This method
is able to detect mutations within splice sites flanking the exons and can therefore detect mutations refractory to analysis by mRNA or cDNA sequencing.

Ornithine transcarbamylase (OTC) deficiency is the most common urea
cycle disorder in man (4) and mutations of the OTC gene tend to occur within

the coding regions, again largely undetectable by Southern analysis. We describe here a method for PCR amplification of OTC cDNA and rapid detection of point mutations by chemical cleavage of mismatches between wild-type and mutant cDNAs and visualization of the cleavage products by polyacrylamide gel eletrophoresis. This is often sufficient to detect and follow a mutation in a particular family, but if more detailed information is required the PCR product can be sequenced directly, as described for the HPRT gene.

Although the methods described here refer to specific genes, the principles and techniques can be applied to any gene for which some sequence information is available, and have universal use in the areas of molecular diagnosis of genetic disease and mutation studies in cultured cells.

2. Direct sequencing of multiplex PCR products: detecting HPRT gene mutations

A general scheme of multiplex PCR amplification is presented in *Figure 1*. Several different regions can be simultaneously amplified *in vitro*, using the PCR, and if the amplified fragments vary in size it is possible to view each of them after separation by gel electrophoresis (5). The coding region of the human HPRT gene is distributed over 39.8 kb DNA and 57 kb of this locus have recently been sequenced (6), yielding the information for the synthesis of oligonucleotide primers suitable for amplification of the individual exons. Although the gene contains nine exons, the seventh and eighth lie so close together (171 bp apart) that they can be amplified in a single fragment. Following *in vitro* amplification of the HPRT exons, the resulting PCR products can be used to generate single-stranded DNA templates suitable for sequencing by seeding a second polymerase reaction in which there is only one oligonucleotide. Each exon is sequenced by the use of a specific priming oligonucleotide and standard dideoxy-terminator sequencing methods (3).

If the PCR and sequencing primers are designed appropriately, it is possible to identify not only the sequence of the individual coding regions, but also some of the sequence immediately flanking the exons. This has the advantage of detecting mutations which affect the splicing of the HPRT mRNA and which may not be detectable by other methods.

The amplification and sequencing of all nine exons of the human HPRT gene requires 25 oligonucleotides, ranging in length from 17 to 28 nucleotides. Details of the location of these oligonucleotides are shown in *Figure 2* together with an indication of the length of the individual PCR products and the relative position of the exons within those amplification units. *Table 1* shows the sequences of each of the PCR and sequencing primers.

2.1 Source material

Since the PCR is such a sensitive procedure, only small quantities of starting

Belinda J. F. Rossiter, Markus Grompe, and C. Thomas Caskey

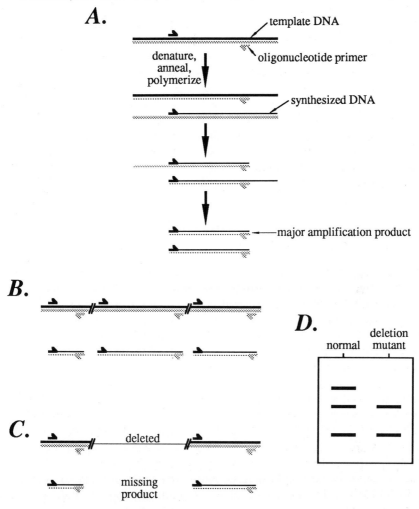

Figure 1. Outline of multiplex PCR strategy.

A. A pair of oligonucleotide primers are synthesized to be complementary to the target sequence, such that each corresponds to opposite strands of the DNA and their 3'-ends face towards each other. Several cycles of separating DNA strands, annealing oligonucleotide primers and DNA polymerization are performed in a thermocycler apparatus, during which time the major amplification product accumulates exponentially since each newly synthesized DNA molecule can act as a template in subsequent rounds of amplification.

B. In the multiplex reaction several different loci can be amplified simultaneously in a single reaction and the differently-sized products distinguished from one another by agarose gel electrophoresis.

C. If a target region in the template DNA is deleted the corresponding amplification product is not generated.

D. Gel electrophoresis of multiplex PCR products can easily demonstrate the deletion of a target region when normal and mutant samples are run alongside each other.

69

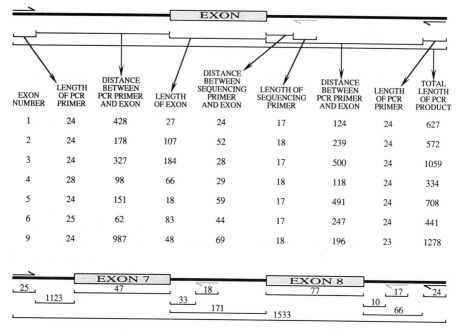

EXON NUMBER	LENGTH OF PCR PRIMER	DISTANCE BETWEEN PCR PRIMER AND EXON	LENGTH OF EXON	DISTANCE BETWEEN SEQUENCING PRIMER AND EXON	LENGTH OF SEQUENCING PRIMER	DISTANCE BETWEEN PCR PRIMER AND EXON	LENGTH OF PCR PRIMER	TOTAL LENGTH OF PCR PRODUCT
1	24	428	27	24	17	124	24	627
2	24	178	107	52	18	239	24	572
3	24	327	184	28	17	500	24	1059
4	28	98	66	29	18	118	24	334
5	24	151	18	59	17	491	24	708
6	25	62	83	44	17	247	24	441
9	24	987	48	69	18	196	23	1278

Figure 2. Summary of HPRT multiplex PCR. The upper part of the figure shows a schematic of an exon within the human HPRT gene flanked by two PCR primers and with a sequencing primer close to its 3'-end. The primer above the line indicates that it corresponds to the sense strand of the sequence; the primers located below the line correspond to the antisense strand. Various distances are shown and their precise sizes displayed in the table below, all sizes are base-pairs. The lower part of the figure shows a schematic of the particular case of exons 7 and 8 since they lie so close together. The appropriate distances are also marked.

material are required. This may be DNA extracted according to *Protocol 1* from cultured cells, blood samples, or small amounts of tissue, such as chorionic villi. The DNA may even be of relatively poor quality (sheared or degraded) or relatively crude (containing residual protein), although in the latter case the PCR mixture should be heated to 94 °C for 10 min before addition of *Taq* polymerase to inactivate any protein present.

Protocol 1. Preparation of genomic DNA starting material

1. Rinse blood cells, harvested lymphoblastoid or fibroblast cells with phosphate buffered saline twice by centrifugation (3000 g/5 min).

2. Resuspend the cells in approximately 2 vol. 10 mM Tris–HCl (pH 7.5), 400 mM NaCl, 2 mM Na₂ EDTA. Add SDS and proteinase K to concentrations of 0.8% (w/v) and 100 μg/ml, respectively; incubate at 55 °C for 2 h.

Table 1. Oligonucleotide primers for the multiplex amplification and direct DNA sequencing of all HPRT exons

All sequences are shown as 5' to 3'. The '5' PCR primers' correspond to the 'sense' strand of the genomic DNA and the '3' PCR primers' and 'sequencing primers' correspond to the opposite strand. The number in parentheses after each PCR primer represents in picomoles the amount required for a 50 μl multiplex PCR amplification as described in *Protocol 2*.

Exon	5'PCR primer	3'PCR primer	sequencing primer
1	TGGGACGTCTGGTCCAAGGATTCA (10)	CCGAACCCGGGAAACTGGCCGCCC (10)	TGGGCCTGAACCGGCCA
2	TGGGATTACACGTGTGAACCAACC (20)	GACTCTGGCTAGAGTTCCTTCTTC (20)	CAGAACAGCTGCTGATGT
3	CCTTATGAAACATGAGGGCAAAGG (16)	TGTGACACAGGCAGACTGTGGATC (16)	ACCTACTGTTGCCACTA
4	TAGCTAGCTAACTTCTCAAATCTTCTAG (25)	ATTAACCTAGACTGCTTCCAAGGG (25)	TGACTCGGTATGAAGTGC
5	CAGGCTTCCAAATCCCAGCAGATG (20)	GGGAACCACATTTTGAGAACCACT (20)	CTGGCTTACCTTTAGGA
6	GACAGTATTGCAGTTATACATGGGG (10)	CCAAAATCCTCTGCCATGCTATTC (10)	AGCAATCACTTAATCCC
7	GATCGCTAGAGCCCAAGAAGTCAAG (18)		CTTTTAGGTTAAAGATGG
8		TATGAGGTGCTGGAAGGAGAAAAC (18)	GAGTGAGAAAAGAAGC
9	GAGGCAGAAGTCCCATGGATGTGT (16)	CCGCCCAAAGGGAACTGATAGTC (16)	GATGGCCACAGAACTAGA

3. Cool to r.t. and add 1/3 vol. saturated NaCl (approx. 6 M), mix gently. Remove precipitated protein and cellular debris by centrifugation (10 000 g/10 min).

4. Add the supernatant to 2 vol. of ethanol at r.t. and recover the DNA by spooling on to a clean glass rod. Rinse with 70% (v/v) ethanol, briefly air-dry, and redissolve in 10 mM Tris–HCl (pH 8.0), 1 mM Na$_2$ EDTA. Prepare a further dilution of the DNA in H$_2$O, at a concentration of 50 μg/ml.

2.2 Design of multiplex PCR primers

The precise locations of a pair of primers to be used for PCR amplification of a single exon do not appear to have very stringent requirements. A multiplex PCR is, however, less forgiving in that several oligonucleotides made for a single exon may not work satisfactorily in the complex reaction. A number of principles are used in the choice of such primers; these may have to be compromised to a certain extent according to the content and length of the intron sequence available.

For subsequent analysis by agarose gel electrophoresis, the PCR products should be at least 150 bp and differ in size by at least 40 bp (or more if the fragments are larger). Oligonucleotides are chosen to be 24 nucleotides long and 50% G/C-rich. If this is not possible a longer A/T-rich or shorter G/C-rich molecule is synthesized so as to have similar melting properties. The priming oligonucleotides should be situated far enough from the exon that the immediate flanking regions can be amplified in addition to the coding sequence. If oligonucleotides are located wholly or partly within the exons, deletions can still be detected but sequencing of the entire exon and flanking regions will not be possible. There should be no spurious hybridization of oligonucleotide primers to non-specific sequences (e.g. repeat sequences) or to other oligonucleotides in the cocktail.

In the multiplex PCR amplification of HPRT gene exons, since the priming oligonucleotides hybridize with intron sequences there is no chance of the primers amplifying cDNA, which may be a contaminant from other experiments in the laboratory, since cDNA molecules lack the genomic intron sequences.

Protocol 2. Multiplex amplification of HPRT gene exons

1. Add to a 0.5 ml microcentrifuge tube:

- H_2O to 45 µl final volume
- 10 µl 5× multiplex reaction buffer[a]
- 3 µl 25 mM each dNTP
- 10–25 pmol of each oligonucleotide primer (see *Table 1* for quantities of HPRT primers)

This reaction mix is stable for up to 3 months at $-70°C$.

2. Mix gently, add 250 ng (5 µl) template DNA and 4 units *Taq* DNA polymerase, mix gently.

3. Add 25 µl paraffin oil, centrifuge briefly (5 sec).

4. Perform PCR in automatic thermocycler: Denature at 94°C for 4.5 min, then 25 cycles of [denaturation (94°C/30 sec), annealing (61°C/50 sec), polymerization (68°C/2 min)], followed by polymerization at 68°C for 5 min. Store at 4°C until analysis (up to 2 months).

5. Electrophorese 15 µl of the reaction product through a 1.4% agarose gel in TBE (90 mM Tris base, 90 mM boric acid, 1 mM Na_2 EDTA). Prepare the gel with 0.5 µg/ml ethidium bromide, or stain after electrophoresis to allow visualization and photography of the results.

[a] 5× multiplex reaction buffer: 83 mM $(NH_4)_2SO_4$, 335 mM Tris–HCl (pH 8.8), 33.5 mM $MgCl_2$, 25 mM 2-mercaptoethanol (βME), 34 µM Na_2 EDTA.

2.3 Single-strand-producing reactions (SSPR)

For each individual exon to be sequenced, a SSPR is performed (see *Protocol 3*) using only one of the primers used in the original PCR reaction. This primer must be of opposite sense to the specific sequencing primer (in the case of HPRT, the '5' PCR primer').

Protocol 3. Generation of single-stranded sequencing templates

1. Add to a 0.5 ml microcentrifuge tube:
   - H_2O to 99.5 µl final vol.
   - 10 µl 5× SSPR buffer[a]
   - 1 µl 25 mM each dNTP
   - 30 pmol primer
   - 0.5 µl multiplex PCR product
   - 5 units *Taq* DNA polymerase.

2. Overlay the reaction with 50 µl paraffin oil, centrifuge briefly (5 sec).

3. Carry out the SSPR under the conditions described in *Protocol 2* (step 4), except that 30 cycles are performed.

4. Dilute the SSPR product with an equal vol. of H_2O, then add the same amount of 7.5 M ammonium acetate and mix. Add 2.5 vol. 100% ethanol and chill for 15 min at −70°C, then spin for 15 min in a microfuge.

5. Take up the pellet in 100 µl H_2O and repeat the ammonium acetate precipitation. Rinse the pellet with 70% (v/v) ethanol, then 100% ethanol, and dry to completion under vacuum. Dissolve in 10 µl H_2O immediately before use in the sequencing reaction (*Protocol 4*).

[a] 5× SSPR buffer: 250 mM KCl, 50 mM Tris–HCl (pH 8.8), 7.5 mM $MgCl_2$.

2.4 Design of sequencing primers

Sequencing primers were constructed to hybridize 10–70 bp downstream of each exon, far enough away that the immediate splice sites flanking the exons could be sequenced, and close enough that the sequence of the entire exon could be determined without too much unnecessary sequence. All the HPRT sequencing primers are 17 or 18 nucleotides long. Sequencing reactions are performed as described in *Protocol 4*.

Protocol 4. Sequencing of multiplex PCR products

1. End-label the sequencing oligonucleotide primer as follows. Mix:
   - H_2O to a final vol. of 50 µl
   - 10 µl 5× kinase buffer[a]
   - 20–50 pmol primer

Protocol 4. *Continued*

- 50–70 μCi [γ-^{32}P] ATP (6000 Ci/mmol)
- 30 units of T4 polynucleotide kinase.

Incubate at 37°C for 45 min.

2. Purify the end-labelled primer by passage through a NENsorb column (NEN/DuPont), dry the product to completion under vacuum and re-suspend in 12 μl H$_2$O immediately before use.

3. Add 5 μl single-strand template (*Protocol 3*) to 3 μl labelled sequencing primer and 2 μl 5× POL buffer.^{b}

4. Heat to 95°C for 10 min and spin briefly (5 sec) in a microfuge to bring down condensation.

5. At r.t. dispense 2.5 μl aliquots of the primer/template mixture into four appropriately labelled tubes (1A, 1C, 1G, 1T, 2A, etc.).

6. Add 2 μl of the appropriate dideoxy-terminator/Sequenase mix [80:8 μM deoxy:dideoxy mix with 0.5 unit/μl Sequenase (United States Biochemical Corp.) added just before use] to each of the four tubes and place them immediately at 50°C for 10 min.

7. Bring down condensation with a brief spin and add 3 μl STOP solution.^{c} Heat to 80°C for 3 min and then analyse by standard sequencing electrophoresis and autoradiography.

^{a} 5× kinase buffer: 50 mM Tris–HCl (pH 7.6), 50 mM MgCl$_2$, 5 mM dithiothreitol (DTT).
^{b} 5× POL buffer: 250 mM Tris–HCl (pH 8.3 at 37°C), 40 mM MgCl$_2$, 150 mM KCl, 50 mM DTT.
^{c} STOP solution: 95% (v/v) formamide, 20 mM Na$_2$ EDTA, 0.05% (w/v) Bromophenol Blue, 0.05% (w/v) xylene cyanol.

2.5 Results and interpretation

Figure 3 illustrates the result of multiplex amplification of the human HPRT gene. The normal gene yields eight amplification products containing the nine exons (exons 7 and 8 are amplified in one fragment because they lie only 171 bp apart). Four cell lines with partial or total gene deletions are included in the analysis and generate from zero to seven PCR products according to the extent of the deletion.

Figure 4 indicates the detection of a point mutation within exon 7 of the HPRT gene from a Lesch–Nyhan patient. Sequencing of the same exon from his mother reveals two different nucleotides in that position, indicating that she is a heterozygote carrier of the mutation. Thus, this method not only has value in detecting mutations within the HPRT genes of affected individuals, but also in genetic analysis of related family members.

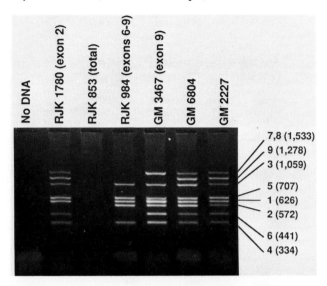

Figure 3. Multiplex amplification of the HPRT locus. All nine exons of the human HPRT gene were amplified on eight separate DNA fragments using 16 oligonucleotide primers in a single PCR reaction. The lane containing no DNA was a control to ensure there was no spurious amplification due to endogenous contaminating sequences. Four examples of complete or partial HPRT gene deletions are shown. The cell lines GM6804 and GM2227 represent the normal multiplex pattern of eight bands. The exons contained within the individual fragments and the size of the amplification units is given on the right. (Reproduced from reference 3 with permission.)

3. Chemical cleavage of PCR products: detecting OTC gene mutations

The majority of mutations within the human OTC gene occur in the coding region. Since this coding region is too long (1062 bp) to routinely sequence in its entirety, a method has been developed to scan PCR-amplified cDNA from OTC mRNA for mismatches relative to the wild-type molecule (7). This scanning method is based on the chemical cleavage method utilizing hydroxylamine and osmium tetroxide (8), and is illustrated in *Figure 5*.

Heteroduplexes are prepared by denaturing and rehybridizing a mixture of wild-type and mutant cDNAs (obtained by PCR amplification; *Protocol 5*). The formation of heteroduplexes is favoured by the use of limiting amounts of wild-type cDNA. One of the strands of the wild-type cDNA is labelled by incorporation of a radiolabelled oligonucleotide primer during the PCR reaction (*Protocol 6*): either the 'sense' or the 'antisense' strand is labelled. Treatment with hydroxylamine results in modification of mismatched C nucleotides and with osmium tetroxide, mismatched T nucleotides. Subsequent exposure of the heteroduplexes to piperidine results in cleavage at

NORMAL   MUTANT HETEROZYGOTE

T C A G   T C A G T C A G

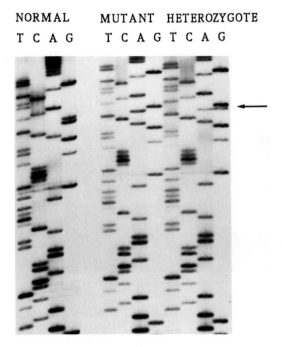

Figure 4. Sequence analysis of wild-type, mutant, and heterozygote HPRT. A G to A transition was identified in Lesch–Nyhan syndrome patient RJK1930 by manual direct DNA sequencing of PCR amplified exon 7 DNA. The mutant allele was also identified in his mother by the presence of two bases at a single position in the autoradiograph (arrow). (Reproduced from reference 3 with permission.)

the modified residues, the products of which are visible by autoradiography if the original labelled end is retained (*Protocol 7*). Thus, by performing four separate reactions for each mutant cDNA, i.e. two labelled strands of the wild-type cDNA and two chemical reactions, it is possible to detect all possible mismatches resulting from point mutations in the cDNA.

3.1 Design of PCR primers

The locations of the amplification primers were chosen so that they would lie outside the coding region of the OTC cDNA. Since the template for the PCR is single-stranded cDNA it is important that the PCR primers be in the correct orientation. Single-stranded cDNA is of the antisense orientation with respect to the mRNA and so the 5′PCR primer must correspond to the sense strand in order to hybridize with the cDNA. The 3′PCR primer, of course, lies on the opposite strand to the 5′PCR primer. The sequences of the PCR primers used for the amplification of OTC cDNA are: '5′PCR primer', TCACTGCAGCTGAACACATTTCTTAG; '3′PCR primer', TGGCA-GATGCAGTATTGGCTCGAGTG.

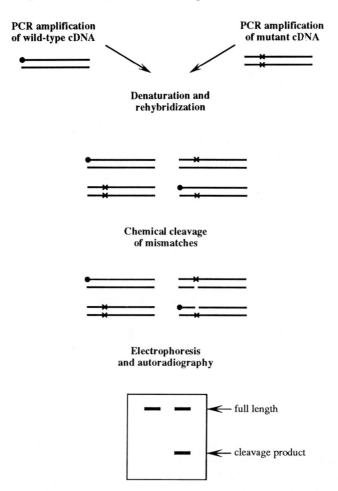

Figure 5. Outline of mismatch chemical cleavage strategy. Templates for the two cDNA species are prepared by PCR amplification. A point mutation within the mutant cDNA is indicated by a cross and one of the wild-type strands is radiolabelled by incorporation of an end-labelled oligonucleotide into the PCR product, indicated by a black circle. On rehybridization, four species are theoretically obtained, two being the original homo-duplexes and the others being heteroduplexes made up of strands from both wild-type and mutant cDNAs. On exposure to the chemicals hydroxylamine or osmium tetroxide, and then piperidine, there may be cleavage at the site of mismatch in the heteroduplexes, after which time the reaction products are separated by electrophoresis. Any DNA strand which retains the end with the originally labelled oligonucleotide is visible after auto-radiography.

3.2 Procedures

3.2.1 cDNA preparation

Protocol 5 describes methods for the synthesis (part A) and subsequent PCR amplification (part B) of wild-type and mutant cDNAs.

Protocol 5. Synthesis and amplification of cDNA

A. *Synthesis of cDNA*

1. Prepare all reagents to be used with RNA with diethylpyrocarbonate-treated H_2O.
2. Mix $10\,\mu g$ total cellular RNA prepared from liver with $4\,\mu g$ oligo-deoxythymidine and take up in $25\,\mu l$ $50\,mM$ Tris–HCl (pH 8.3), $8\,mM$ $MgCl_2$, $30\,mM$ KC1.
3. Heat to 65 °C for 3 min, chill on ice, and centrifuge briefly (5 sec) to bring down condensation.
4. Add:
   - 14 units RNA guard (Pharmacia LKB Biotechnology)
   - $3\,\mu l$ $100\,mM$ DTT
   - $1.5\,\mu l$ $25\,mM$ each dNTP
   - 100 units Moloney murine leukaemia virus reverse transcriptase (total vol. $30\,\mu l$).

   Incubate at 42 °C for 1 h.
5. Hydrolyse RNA with $45\,\mu l$ $0.7\,M$ NaOH, $40\,mM$ Na_2 EDTA and heat to 65 °C for 10 min.
6. Precipitate single-stranded cDNA with the addition of $8\,\mu l$ $2\,M$ ammonium acetate (pH 4.5) and $200\,\mu l$ ethanol, chill at -20 °C for at least 4 h and preferably overnight. Spin in a microcentrifuge, 10 min, rinse pellet with 70% (v/v) ethanol, then 100% ethanol, and dry under vacuum.

B. *Amplification of cDNA by PCR*

7. Take up single-strand cDNA in $20\,\mu l$ H_2O and use $10\,\mu l$ for the PCR reaction.
8. Add to the cDNA in a 0.5 ml microcentrifuge tube:
   - H_2O to $99\,\mu l$ final vol.
   - $20\,\mu l$ 5× reaction buffer[a]
   - $6\,\mu l$ $25\,mM$ each dNTP
   - $10\,\mu l$ DMSO
   - 50 pmol of each oligonucleotide primer

   Add 5 units *Taq* DNA polymerase, mix gently, add $25\,\mu l$ paraffin oil, centrifuge briefly (5 sec).

9. Perform PCR in automatic thermocycler: Denature at 95°C for 7 min, then 30 cycles of [denaturation (90°C/30 sec), annealing (45°C/60 sec), polymerization (67°C/3.5 min)]. Store at 4°C.

10. Adjust volume of PCR reaction to 200 µl with H_2O, add 60 µl 10 M ammonium acetate, mix and spin in a microcentrifuge for 20 min.

11. Carefully transfer supernatant to fresh tube and add 600 µl ice-cold ethanol, chill at −70°C for 15 min and spin in a microcentrifuge for 15 min.

12. Dissolve pellet in 200 µl H_2O and repeat steps 10 and 11.

13. Having removed unincorporated nucleotides and oligonucleotide primers by the two ammonium acetate precipitations, quantitate DNA. Dilute to a concentration of 1 ng/µl (wild-type cDNA) or 50 ng/µl (mutant cDNA) in H_2O and store at −20°C.

[a] 5× reaction buffer: 83 mM $(NH_4)_2SO_4$ 335 mM Tris–HCl (pH 8.8), 33.5 mM $MgCl_2$ 25 mM βME, 34 µM Na_2 EDTA, 850 µg/ml bovine serum albumin.

3.2.2 Generation of wild-type and mutant heteroduplexes

Mutant and wild-type cDNAs are mixed and annealed under conditions favouring heteroduplex formation. Subsequent PCR amplification in the presence of suitable end-labelled primers yields substrates for chemical cleavage reactions.

Protocol 6. Generation of wild-type and mutant heteroduplexes

1. End-label 3.5 pmol of each PCR oligonucleotide primer in 30 µl as described in *Protocol 4*, step 1 and purify through NENsorb columns (NEN/DuPont).

2. Set up two 50 µl PCR reactions, each with one labelled primer and one unlabelled, and with 1 ng of wild-type PCR product as template. Use the same reaction mixture and conditions described in *Protocol 5*, steps 7 and 8, but with only 2.5 pmol of each primer and 25 rounds of amplification. Since the oligonucleotide primers are limiting under these conditions, this should yield approximately 2.5 pmol product.

3. Isolate the labelled PCR products from low-melting-point agarose and determine the specific activity (approximately 10^7 c.p.m./µg). These probes may only be kept for 2 to 3 weeks before radiolysis renders them impractical.

4. Mix 10 ng of probe (approximately 100 000 c.p.m.) with 150 ng unlabelled (mutant) target DNA in 30 µl 0.3 M NaCl, 3.5 mM $MgCl_2$, 3 mM Tris–HCl (pH 7.7) [use a 5× stock buffer] and boil for 5 min.

5. Chill on ice for 5 min and allow rehybridization at 42°C for 2 h.

Protocol 6. *Continued*

6. Add 90 µl ice-cold ethanol and precipitate at −70°C for 15 min. Spin down heteroduplexes in a microcentrifuge, 10 min, rinse with 70% (v/v) ethanol and resuspend in 20 µl H_2O.

3.2.3. Chemical cleavage

The times given here have been shown to be optimal for the analysis of OTC cDNA, but it may be necessary to modify these in other cases. The initial use of a time-course experiment is recommended. Prepare the following reaction mixtures:

(a) *Hydroxylamine solution.* Dissolve 1.39 g hydroxylamine chloride in 1.6 ml H_2O under warm water, add approximately 1 ml diethylamine, further adjust the pH to 6.0 with diethylamine. Estimate the pH of the hydroxylamine stock by adding a few drops to H_2O and measuring the pH of that portion. The solution may be kept at 4°C for 7 to 10 days.

(b) *Osmium tetroxide.* Prepare a 2% stock by dilution of a commercial 4% stock, discard when the solution becomes discoloured (about 3 months). Store at 4°C.

(c) *HOT STOP buffer* (0.3 M sodium acetate (pH 5.2), 0.5 mM Na_2 EDTA, 25 µg/ml tRNA). Autoclave sodium acetate/Na_2 EDTA stock, add tRNA when required to 10 ml aliquots. Clean tRNA before use by phenol and chloroform extractions, sodium acetate/ethanol precipitation, dissolution in H_2O and heat treatment at 95°C for 10 min. Store at −20°C.

Protocol 7. Chemical cleavage reactions

1. Set up 4 cleavage reactions for each PCR produce, i.e. one hydroxylamine and one osmium tetroxide reaction for each strand. Dispense 6 µl of the appropriate heteroduplex into each reaction tube, place the osmium tetroxide tubes on ice.

2A. **Hydroxylamine reactions**: Add 20 µl hydroxylamine solution [see 3.2.3(a)], mix well and incubate at 37°C for 20 min. Add 200 µl HOT STOP buffer [see 3.2.3.(c)] and 750 µl ice-cold ethanol, chill at −70°C for 20 min and spin in a microcentrifuge for 20 min. Rinse with 70% (v/v) ethanol and dry under vacuum.

2B. **Osmium tetroxide reactions**: Add on ice 2.5 µl 10× osmium tetroxide buffer.[a] and 15 µl 2% osmium tetroxide [see 3.2.3(b)], mix well (yellow colour reaction) and incubate at 37°C for 5 min with frequent agitation. Transfer back to ice, add 200 µl HOT STOP buffer [see 3.2.3(c)] and 750 µl ice-cold ethanol, chill at −70°C for 20 min and spin in a microcentrifuge for 20 min. Rinse with 70% (v/v) ethanol and dry under vacuum.

3. **Piperidine cleavage**: To each tube add 50 μl 1 M piperidine and mix, incubate at 90 °C for 30 min. Chill on ice, then add 50 μl 0.6 M sodium acetate (pH 5.2) and 250 μl ice-cold ethanol. Chill at −70 °C for 20 min then spin in a microcentrifuge for 20 min (if possible use a microcentrifuge that pellets to the bottom of the tube), rinse with 70% (v/v) ethanol and dry under vacuum.

4. Redissolve pellets in 5–15 μl 100% formamide *thoroughly*. Check with a Geiger counter that at least 50% of the counts are in the liquid. Heat to 90 °C for 3 min and chill on ice.

5. Analyse the samples on a 4% denaturing polyacrylamide gel, followed by autoradiography. A wedge gel gives resolution of fragment sizes from 30 to 1200 bp; the gel should be stopped when the first dye (Bromophenol Blue) reaches the bottom.

a 10× Osmium tetroxide buffer: 100 mM Tris–HCl (pH 7.7), 10 mM Na_2 EDTA, 15% (v/v) pyridine.

3.3 Results and interpretation

The results of chemical cleavage of OTC cDNA from five patients suffering from OTC deficiency is shown in *Figure 6*. Each patient has a different cleavage pattern, indicating that the mutations are all different. By performing four reactions with each sample (either strand of the wild-type cDNA labelled and two chemical reactions) it is possible to identify all possible point mutations and to determine their approximate location. This technique has been used to detect six OTC mutations studied in our laboratory. In most cases this analysis is sufficient to detect a mutation in an affected individual and in family members who may be carriers. It is, however, quite straightforward to amplify the appropriate region using the PCR, and to sequence the mutant allele if that is required; the methods described for HPRT gene exon sequencing can be used for such analysis. For direct sequencing of *in vitro* amplified OTC DNA, either cDNA or genomic DNA could be used as the original targets.

Acknowledgements

B.J.F.R. is a Fellow of the Arthritis Foundation, M.G. is an Association of Medical School Pediatric Department Chairman, Inc. Pediatric Scientist Training Program Fellow supported by NIH Grant #00850, and C. T. C. is an Investigator with the Howard Hughes Medical Institute.

Detection of deletions and point mutations

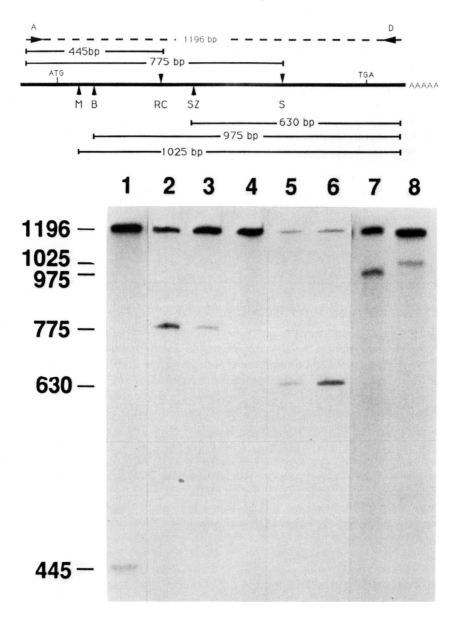

Figure 6. Chemical cleavage of OTC cDNA in five patients. A schematic representation of mismatch sites and cleavage fragments in the OTC cDNA is shown in the upper part of the figure. Sites of putative mutations in patients M, BB, RC, SZ, and S are indicated by arrows. Cleavage products (and their length in base-pairs) are indicated by the five thin lines. The dashed line indicates the extent of the original PCR product, between primers A and D. The lower part of the figure shows an autoradiograph in which cleavage products can be observed. The 'sense' probe was used in **lanes 1–4** and the 'antisense' probe in **lanes 5–8**. **Lanes**: (**1**) patient RC (osmium tetroxide cleavage), (**2**) and (**3**) patient S (hydroxylamine and osmium tetroxide cleavage), (**4**) wild-type control (no cleavage), (**5**) and (**6**) patient SZ (hydroxylamine and osmium tetroxide cleavage), (**7**) patient BB (osmium tetroxide cleavage), (**8**) patient M (hydroxylamine cleavage). (Reproduced from ref. 7 with permission.)

References

1. Mullis, K. B. and Faloona, F. A. (1987). In *Methods in enzymology* (ed. R. Wu). Academic Press, London and New York. 155, pp. 335–50.
2. Stout, J. T. and Caskey, C. T. (1989). In *The metabolic basis of inherited disease* (ed. C. R. Scriver, A. L. Beaudet, W. S. Sly, and D. Valle,) pp. 1007–28. McGraw-Hill, New York.
3. Gibbs, R. A., Nguyen, P. N., Edwards, A., Civitello, A. B., and Caskey, C. T. (1990). *Genomics*, **7**, 235.
4. Brusilow, S. W. and Horwich, A. L. (1989). In *The metabolic basis of inherited disease* (ed. C. R. Scriver, A. L. Beaudet, W. S. Sly, and D. Valle), pp. 629–63. McGraw-Hill, New York.
5. Chamberlain, J. S., Gibbs, R. A., Ranier, J. E., Nguyen, P. N., and Caskey, C. T. (1988). *Nucleic Acids Research*, **16**, 11141.
6. Edwards, A., Voss, H., Rice, P., Civitello, A., Stegemann, J., Schwager, C., Zimmermann, J., Erfle, H., Caskey, C. T., and Ansorge, W. (1990) *Genomics*, **6**, 593.
7. Grompe, M., Muzny, D. M., and Caskey, C. T. (1989). *Proceedings of the National Academy of Sciences, USA*, **86**, 5888.
8. Cotton, R. G. H., Rodrigues, N. R., and Campbell, R. D. (1988). *Proceedings of the National Academy of Sciences, USA*, **85**, 4397.

6

PCR of TG microsatellites

MICHAEL LITT

1. Introduction

It is less than 10 years since the proposal was made to construct a genetic linkage map of man using restriction fragment length polymorphisms (RFLPs) (1). Since that time more than 3000 RFLPs have been described (2) and linkage maps have been published for many human chromosomes. Using these tools, numerous Mendelian diseases of man have been genetically mapped and it is generally agreed that, in principle, any human Mendelian trait may now be localized. However, certain practical considerations limit the application of linkage mapping in man. One of these considerations is the limited polymorphism information content (PIC) (1) of most of the known marker loci. Although the discovery of minisatellites, also known as variable number of tandem repeat (VNTR) polymorphisms (3, 4) has helped to reduce this limitation, highly informative VNTR loci have not been found on all chromosome arms and those which have been identified are often situated near telomeres (5), leaving large regions of the genome out of reach of multi-allelic marker loci. Highly polymorphic marker loci may also be constructed by haplotyping several closely linked site polymorphisms. Although linkage disequilibrium may decrease the PIC of such compound loci, numerous examples exist for which PICs exceed 0.70. A disadvantage of compound loci is that individuals who are multiply heterozygous cannot always be haplotyped. Due to the fact that sibships are often small and parents of affected individuals are usually unavailable, this problem is especially severe when compound loci are used for affected sib-pair (6) or affected relative-pair (7) analyses of late-onset diseases, such as familial Alzheimer's disease. In such cases, VNTR loci are especially useful, because their full information content is available without haplotyping. Given a sufficiently dense set of VNTRs, disease genes may be mappable by determining the degree of sharing alleles at numerous marker loci. In affected relative pairs, marker loci closely linked to a disease gene will share alleles more frequently than expected by chance.

The human genome contains approximately 50–100 000 copies of an interspersed repeat with the sequence (dT-dG/dA-dC)n, where n = approximately 10 to 60. (See references cited in (8).) In humans, TG/AC repeats ('TG microsatellites') have been found in several sequenced regions. Since mini-satellite regions with larger repeat elements often display extensive length polymorphisms, it seemed possible that TG microsatellites might also be polymorphic. Recent evidence indicates that this is true. Weber and May (9) screened the GenBank database and found several TG microsatellite-containing DNA sequences that had been sequenced independently by at least two different laboratories. In seven such cases, the reported sequences had different numbers of dinucleotide repeats. Using the polymerase chain reaction (PCR) to amplify several of these TG microsatellites, and separating the products on DNA sequencing gels, they confirmed the polymorphic nature of several such loci and showed that they followed Mendelian segrega-tion in families. Independently, we characterized TG microsatellite VNTRs within an intron of the cardiac actin gene (8) and within an arbitrary DNA segment, D11S35, on the long arm of chromosome 11 (10). As expected for VNTR polymorphisms involving dinucleotide repeats, allelic fragments dif-fered in length by multiples of two nucleotides. *Table 1* lists 44 human TG microsatellites in order of decreasing PIC and shows their regional localiza-tions, where known. The average PIC of these loci is 0.57 and 36% of them have PICs ⩾0.70. In contrast to minisatellite-related VNTRs, or restriction site polymorphisms (11), highly polymorphic TG microsatellites appear to be abundant on the X chromosome (12). Also in contrast to minisatellite VNTRs, the available data for TG microsatellites show no tendency for them to cluster near chromosomal telomeres.

Several published reports (13, 14) have described other types of simple sequence polymorphisms in human or other eukaryotic genomes. However, because TG microsatellites appear to be the most abundant class of dispersed simple sequence repeats, we have focused our efforts on the characterization of polymorphisms based on variation involving this class of simple sequence.

2. Ascertaining and scoring microsatellite VNTRs

2.1 Screening for TG microsatellites

Due to their abundance and wide distribution, TG microsatellites may be found by screening any type of genomic library. However, to minimize the number of clones which need to be screened, it is helpful to use genomic libraries with large inserts. Approximately 40% of cosmids from a random human genomic library contained at least one TG microsatellite (8); similar results have been found with cosmid libraries specific for human chromo-somes 2, 17, 21, and 11 q (our unpublished results). In large insert bacterio-phage libraries of total human genomic DNA (15) or flow-sorted human X chromosomes (12) approximately 20–25% of the clones contain TG micro-satellites. When searching for a TG microsatellite closely linked to a previously cloned single-copy DNA segment, it is useful to screen a cosmid library with

Table 1. Some characteristics of 42 (TG/CA) microsatellites

Locus symbol	PIC	Chromosomal localization	Reference
ACTC	0.86	15q	8
ASS	0.83	9q32-q34	D. J. Kwiatkowski, pers. comm.
Mfd15	0.81	?	2, 27
APOC2	0.79	19q12-q13.2	9
D19S47	0.79	19	2, 27
DXS424	0.79	Xq24-q26	12
DXS425	0.79	Xq26-q27.1	12
D11S35	0.78	11q22	10
Mfd26	0.78	?	2,27
D21S167	0.78	21q22.2	31
DXS456	0.77	Xq21-q22	12
GSN	0.76	9q32-q34	D. J. Kwiatkowski, pers. comm.
Mfd23	0.75	?	2, 27
Mfd14	0.74	?	2, 27
D19S49	0.71	19	2, 27
DRD2	0.70	11q22-q23	X. Y. Luo, pers. comm.
Mfd18	0.69	?	2, 27
Mfd19	0.68	?	2, 27
ERCC1	0.66	19q13.2-q13.3	B. Wieringa, pers. comm.
APOA2	0.65	1q21-q23	9
Mfd32	0.65	?	2, 27
Mfd25	0.64	?	2, 27
W30	0.60	7q21-q31	M. Dean, pers. comm.
Mfd8	0.58	?	2, 27
Mfd20	0.53	?	2, 27
DXS426	0.52	Xp11.21-p21.1	12
HMG14	0.50	21	S. Antonarakis, pers. comm.
Mfd6	0.50	?	9
SST	0.46	?	9
PENK	0.43	8q23-q24	2, 27
Mfd12	0.43	?	2, 27
D19S48	0.42	?	2, 27
R-Ras	0.36	19q13.3	2, 27
RHO	0.31	3q21-q24	9
PDX17-1	0.30	17	J. A. Luty, pers. comm.
DXS457	0.20	Xq21.2-q22	12
PDHA-B	0.19	Xq22	H. Dahl, pers. comm.
PDHA-C	0.14	Xq22	H. Dahl, pers. comm.
PDHA-A	0.09	Xq22	H. Dahl, pers. comm.
PDX17-2	0.00	17	J. A. Luty, pers. comm.
D11S286	0.00	11q24	J. A. Luty, pers. comm.
C21G-10A	0.00	21	Z. Guo, pers. comm.

the single-copy segment and also with a probe homologous to the micro-satellite, in the hope that one or more cosmids will hybridize to both probes. If the screening does not yield such a clone, it should often be obtainable by a few steps of chromosome walking.

To screen libraries for microsatellite repeats, we have successfully used both nick translated poly (dT-dG)·poly(dA-dC) and a 5′-end labelled (dA-dC)$_{10}$ oligomer. Although in theory the oligomer probe is expected to be more specific than the alternating copolymer, both probes appear to give satisfactory results (see *Protocol 1*).

Protocol 1. Screening libraries for TG/CA repeats with

A. *Poly (dT-dG)·poly(dA-dC) probe*

1. Pre-hybridize at 65 °C in 0.5 M sodium phosphate, pH 7.0, 7% (w/v) SDS, 1% bovine serum albumin (23) for 0.5–2 h.
2. Hybridize colony filters (24) or plaque lifts (25) overnight in the same buffer containing 10^6 d.p.m./ml nick translated (> 10^8 d.p.m./μg) poly (dT-dG)·poly(dA-dC; Pharmacia/LKB No. 27-7940-01) probe.
3. Wash filters at r.t. in 2 × SSC/0.1% SDS 15 min, 0.5 × SSC/0.1% SDS 15 min, and then twice at 55 °C in 0.1 × SSC/0.1% SDS for 15 min.
4. Autoradiograph filters overnight at −70 °C with an intensifying screen.

B. *(dC-dA)$_{10}$ oligomer probe*

1. Obtain or synthesize a sample of the oligomer (dC-dA)$_{10}$ (see *Protocol 2*).
2. Label the oligomer with ^{32}P at its 5′-end by treating with [γ-^{32}P]ATP (6000 Ci/mmol; New England Nuclear) in the presence of polynucleotide kinase (16). Remove unincorporated [γ-^{32}P]ATP from the oligomer by passing over a NENsorb^{TM} column (New England Nuclear).
3. Pre-hybridize for 3 h at 50 °C in 5 × SSPE, 0.1% SDS, 100 μg/ml sonicated denatured salmon sperm DNA.
4. Add 10^6 d.p.m./ml end-labelled oligomer probe and hybridize filters for 3 h.
5. Wash the filters in 6 × SSC/0.1% SDS, first at r.t. four times for 15 min each, then for 5 min at 55 °C, and finally for 30 min at r.t.
6. Autoradiograph the filters overnight at room temperature without intensifying screens.

2.2 Sequencing strategy

Aliquots of cosmid or phage clones believed to contain TG microsatellites are digested with several different 4 base cutter restriction enzymes including *Sau*3A, *Hae*III and *Alu*I. Double digests with *Hae*III and *Alu*I are also performed. Southern blots of such digests are probed with the alternating copolymer to reveal microsatellite containing fragments (see *Figure 1*). Digests with microsatellite-containing fragments of sizes appropriate for DNA sequencing (usually 200–800 bp) are used for subcloning into an

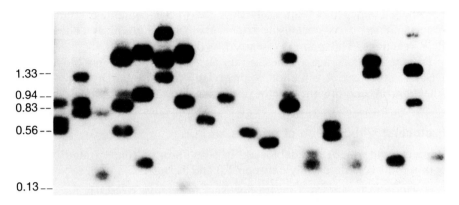

1.33 --

0.94 --
0.83 --

0.56 --

0.13 --

Figure 1. Southern blot of restriction digests of six human genomic cosmids probed with poly(dT-dG).poly(dA-dC). **Lanes 1–3:** *Sau*3A, *Hae*III and *Alu*I/*Hae*III digest of cosmid 1B10; **lanes 4–7:** *Sau*3A, *Alu*I, *Hae*III and *Alu*I/*Hae*III digests of cosmid 1C10; **lanes 8–11** *Sau*3A, *Alu*I, *Hae*III and *Alu*I/*Hae*III digests of cosmid 1D9; **lanes 12–15:** *Sau*3A, *Alu*I, *Hae*III and *Alu*I/*Hae*III digest of cosmid 1F6; **lanes 16–19:** *Sau*3A, *Alu*I, *Hae*III and *Alu*I/*Hae*III digests of cosmid 1H9. Positions of *Eco*RI/*Hind*III fragments of lambda DNA used as size markers are indicated on the left. Conditions for hybridization and washing of the blot were those given in *Protocol 1*.

appropriate plasmid vector such as pTZ18u (US Biochemical Co.). After transformation into a competent bacterial host, subclones are again screened with the poly (dG-dT)·poly (dA-dC) probe, plasmids from positive subclones are prepared as minipreps and sequenced using a dideoxy protocol (16).

It has been suggested that (TG/AC)n repeat blocks with $n \leq 12$ are unlikely to be polymorphic (27). Hence, only microsatellites with $n > 12$ are screened for polymorphism as described below.

2.3 Design of PCR primers

In designing PCR primers for microsatellite amplification, we have followed the recommendations of Saiki *et al.* (17). We have generally used 20-mers with base compositions in the range of 35–55% GC while trying to match the %GC of the two primers to within 5%. It may be possible to compensate for greater differences in the % GC by making minor adjustments in the lengths of the primers thus allowing greater flexibility in their placement. We routinely check primers to insure that they lack complementarity to each other and also that they lack homology with an *Alu* consensus sequence (18). In addition, we try to avoid situating primers in areas containing simple sequence repeats.

Subject to the above constraints, it is advantageous to place the primers as close to the microsatellite repeat as possible. In this way, the length of allelic fragments is minimized, allowing scoring to be performed more quickly and reliably. In addition, the PCR product consists mostly of the dinucleotide repeat, with little flanking sequence, allowing the (CA)n-containing strand to be labelled much more intensely than the (TG)n-containing strand if the single labelled deoxynucleoside triphosphate [α-^{32}P]dCTP is added to the PCR reactions. This point is discussed further below.

Primers for PCR are synthesized on an Applied Biosystems Model 380A DNA Synthesizer or obtained commercially from Research Genetics, Inc. The work-up and quantitation of primers is described in *Protocol 2*.

Protocol 2. Preparation of primers for PCR

Primers (unpurified) are deblocked in a solution of concentrated NH$_4$OH with a total volume usually between 1.0 and 1.5 ml.

1. Evaporate to dryness (Savant Speevac) to remove NH$_4$OH.
2. Resuspend primers in 0.4 ml 10 mM Tris–HCl, 1 mM EDTA, pH 8.0. There may be some insoluble impurities but the primers themselves will dissolve quickly.
3. Extract the suspension with an equal volume of phenol, followed by phenol/CHCl$_3$ (1:1) and finally by CHCl$_3$. Each time, spin the mixture briefly in a microcentrifuge and transfer the clear supernatant to a fresh tube, being careful to avoid inclusion of interface material.
4. Evaporate the final supernatant to dryness to remove CHCl$_3$. Resuspend the final residue in 200 µl deionized water.
5. Read the UV absorbance of a 1:100–1:500 dilution in 10 mM Tris–HCl, 0.1 mM EDTA, pH 8.0, at 230, 260, 280 nm. Calculate the concentration of oligomer assuming 1 mg/ml has an A$_{260}$ = 33;[a] also assume that 1 nmol of a 20-mer = 6.6 µg. From this we can calculate that a 1 nmol/ml solution of a 20-mer will have an A$_{260}$ = 0.166. The 260/230 ratio should be >2.0; the 260/280 ratio should be >1.5.
6. Dilute the stock solution to 100 pmol/µl. Make a working stock solution of each primer by diluting an aliquot of the main stock 1:10 with deionized water, to a final concentration of 10 pmol/µl.
7. Store all primer stocks frozen at −20°C.

[a] The oligomer concentration calculated here is approximate. For more precise quantitation, see ref. 26.

2.4 PCR amplifications

2.4.1 Thermal cycling parameters

As a starting point, the general recommendations of Saiki (17) are useful. However, amplification of microsatellites may be especially demanding in

Michael Litt

that scoring of allelic fragments can often be severely hampered by appearance of spurious bands due to cross-hybridization of primers to non-target sequences. In such cases, it is important to use annealing temperatures that are as high as possible and to use primer concentrations in the range of 0.1 to 0.2 μM, instead of the 1 μM concentration recommended by Saiki *et al.* (17). The effect of annealing stringency is illustrated in *Figure 2*. A typical procedure for trial PCR of a microsatellite locus is given in *Protocol 3*.

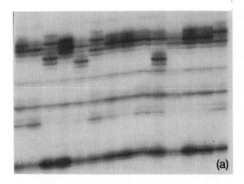 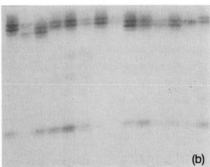

Figure 2. Autoradiographs of gels showing PCR amplified fragments from the W30 locus on chromosome 7q, closely linked to the cystic fibrosis gene (29). The templates used were genomic DNA samples from Utah family K1331. PCR was performed according to *Protocol 3*, with the addition of a single 5'-end-labelled primer. (a) Annealing temperature = 55°C. (b) Annealing temperature = 60°C.

Protocol 3. Trial amplification of genomic microsatellite repeats

1. Prepare a pre-mix containing all of the required components for the reaction (see *Table 2*) except template DNA. The volumes of the components added to the pre-mix shown below are per reaction tube; multiply these volumes by the number of tubes to be run and add 5% to allow for pipetting errors.
   - Sterile H₂O   to final vol. of 20 μl
   - 10 × reaction buffer   2.5 μl
   - dNTP mix   4.0 μl
   - Primers   5.0–25 pmol each
   - *Taq* polymerase   0.625 units
2. Vortex **gently** or pipette up and down to mix.
3. Into 0.5 ml microcentrifuge tubes (autoclaved), pipette 5 μl (~ 25 ng) aliquots of the genomic DNA samples to be amplified.
4. Pipette 20 μl pre-mix into each tube.

91

Table 2. Materials for PCR amplification of genomic (TG)n microsatellite repeats

(a) 10 × reaction buffer:
  - 500 mM KCI
  - 100 mM Tris–HCl, pH 8.3 (at 25 °C)
  - 15 mM MgCl$_2$
  - 0.1% autoclaved gelatin

(b) dNTP mix[a]
  - 1.25 mM each dATP, dCTP, dGTP, TTP

(c) Template:
  - Dilute genomic DNA with water to approximately 5–10 μg/ml.
  - Heat the genomic DNA in a 95–100 °C water bath for 5 min.
  - Cool the solution in ice-water and store it frozen at −20 °C

(d) *Taq* polymerase:
  - From US Biochemical or Perkin-Elmer/Cetus.
  - Treat like a restriction enzyme, except vortex well for about 2 sec just before use.

[a] When using [α-^{32}P] dCTP for labelling PCR products, decrease the concentration of unlabelled dCTP in the dNTP mix to 0.125 mM.

Protocol 3. *Continued*

5. Vortex gently to mix.

6. Pipette 25 μl light mineral oil (Sigma M3516) into each tube and close tightly.

7. Perform 30 cycles of PCR, either manually or with an automated thermal cycler. A typical cycle consists of 1 min at 94 °C, 2 min at the annealing temperature (T_a; varies from 45 °C to 65 °C, depending on the locus being amplified) and 30 sec at 72 °C.

8. At the end of the final cycle, increase the extension (72 °C) step by an additional 4.5 min.

9. Cool the samples to 4 °C and store until further use.

2.4.2 Detection of products

Amplification of each new microsatellite locus is first optimized in trial reactions using non-radioactive reagents (see *Protocol 3*), and the products are analysed on ethidium bromide-stained 1.5–2% agarose minigels to ensure that efficient production of a fragment of the predicted length has been obtained. Microsatellites which meet this test are then amplified in the presence of either a single labelled deoxynucleoside triphosphate such as [α-^{32}P]dCTP (for internal labelling) or a single [5'-^{32}P]-end-labelled primer (end labelling). For internal labelling, use the conditions given in *Protocol 3* but add to the pre-mix 0.5–1.0 µCi [α-^{32}P]dCTP or dATP per reaction tube and decrease the concentration of the corresponding unlabelled dNTP by tenfold. For end labelling, use the conditions given in *Protocol 3* but add to the pre-mix sufficient quantity of a single [5′-^{32}P]-end-labelled primer such that the specific radioactivity of this primer is at least 2×10^4 d.p.m./ pmol. For end-labelling primers, we use the protocol given in ref. 16 (p. 11.31).

Resolve radioactive PCR products on 0.4 mm-thick 5–6% denaturing acrylamide gels of the type used for DNA sequencing (16). A DNA sequence ladder of a known sequence serves as a size standard. Gels are lifted from the glass plate on to Whatman 3MM paper, covered with Saran wrap and dried under vacuum at 80 °C without fixing. Dried gels are autoradiographed with or without intensifying screens for 1–2 days.

2.4.3 Instrumentation

To be useful for scoring microsatellite alleles, thermal cyclers need to provide uniform temperatures at all phases of the PCR cycle. This is especially important when a GC rich segment is being amplified, because the high melting temperature required may be uncomfortably close to the denaturation temperature of *Taq* polymerase. Uniform annealing temperatures are also important to allow elimination of 'background bands' caused by cross-hybridization of primers to non-target sequences. If elimination of these bands is attempted by increasing the stringency of annealing, the existence of temperature gradients within the cycler may give rise to failures of amplification in the hottest parts of the device. In our experience, some commercially available instruments which have proved satisfactory for other applications of PCR are unsuitable for microsatellite work because of well-to-well temperature variations exceeding 5 °C.

We evaluate a thermal cycler for uniformity by using an electronic thermometer (such as Cole-Parmer N-08528-10) equipped with a microprobe (such as Cole-Parmer N-08516-90) to measure the temperature of a dummy reaction tube placed at various positions in the machine. The dummy tube contains the same quantities of aqueous solution and mineral oil as an authentic reaction tube, and the probe is inserted through a small hole punched in the tube cap with a syringe needle. To minimize heat loss from the tube, it is important to perform this test with both the cap of the reaction tube and the cover of the instrument closed.

We have successfully used thermal cyclers manufactured by MJ Research and by Coy Laboratory Products. In our experience, these instruments have well-to-well temperature variations of less than 1 °C during all phases of the PCR cycle. In addition, they produce approximately uniform yields of product when aliquots of the same reaction mixture are amplified in different wells, such that all of the wells are usable. However, both of these devices use Peltier devices for heating and cooling the heating blocks. Since it has been claimed that such devices require frequent replacement (2), prospective purchasers of such instruments would be well-advised to scrutinize carefully the warranties and service contracts available.

2.4.4 Extra bands

In theory, each allele of a target microsatellite sequence should give rise to a single band on a sequencing gel after PCR. However, even in cases where the primers hybridize only to target sequences, PCR of microsatellites usually gives rise to 'extra bands'. Three different mechanisms have been suggested to account for this phenomenon.

The first mechanism, which is relevant only when the products are internally labelled, is due to separation of the (CA)n and (TG)n strands on the sequencing gels (9). If the products are labelled internally with a single radioactive nucleotide, and if they differ sufficiently in their content of that nucleotide, strand separation causes minimal interference with scoring of alleles. An example is shown in *Figure 3B*, which illustrates scoring of the locus D11S35 in 18 unrelated individuals. To obtain this result, PCR was performed manually for 37 cycles using the conditions described in *Protocol 3*. Primers used (each at 1 μM final concentration) were 780 (corresponding to residues 7–26 of the target sequence shown in *Figure 3a*) and 781 (corresponding to the complement of residues 153–172). Each cycle consisted of 1 min at 93°C, 2 min at 40°C and 2 min at 72°C. The PCR products were labelled internally with [α-^{32}P]dCTP. From the target sequence (*Figure 3a*), it can be calculated that the ratio of C content in the (CA)n strand to that in the (TG)n strand is approximately 3:1. Hence the faster moving (CA)n strand is labelled much more intensely than the lagging (TG)n strand.

If the primers immediately flank the microsatellite repeat, the ratio of cytosine content in the (CA)n strand to that in the (TG)n strand will usually be large enough to allow use of internal labelling. However, because TG microsatellite repeats often occur in close association with other simple sequence repeats (9) the constraints on primer design discussed in Section 2.3 often force one to locate the primers at some distance from the (TG/CA) repeat. In such instances the bias in base composition often tends to be lost, and both strands are labelled with comparable intensities. Therefore, internal labelling can give rise to two closely spaced bands per allele with similar intensities, making scoring of alleles difficult. This problem is exacerbated by the presence of extra bands produced by additional mechanisms.

The second mechanism, which also may be relevant only when the products are internally labelled, is the template-independent addition of one or more nucleotides to the 3'-end of the product strands (19, 28). This gives rise to a ladder of fragments separated from the major bands by one-nucleotide spacing. We concur with the observation that this one-nucleotide ladder is generally seen with internal labelling and is usually absent if a single end-labelled primer is used for labelling PCR products (19). It has been suggested (19) that this may

Michael Litt

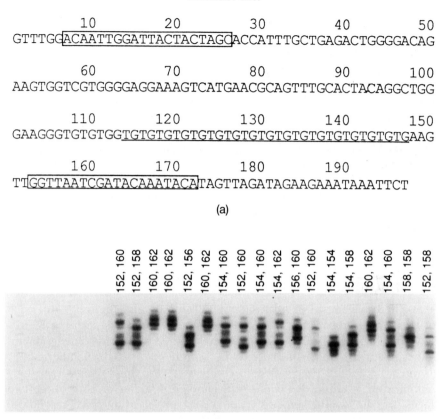

```
          10        20        30        40        50
GTTTGGACAATTGGATTACTACTAGCACCATTTGCTGAGACTGGGGACAG

          60        70        80        90        100
AAGTGGTCGTGGGGAGGAAAGTCATGAACGCAGTTTGCACTACAGGCTGG

          110       120       130       140       150
GAAGGGTGTGTGGTGTGTGTGTGTGTGTGTGTGTGTGTGTGTGTGTGTGAAG

          160       170       180       190
TTGGTTAATCGATACAAATACATAGTTAGATAGAAGAAATAAATTCT
```

(a)

(b)

Figure 3. (a) DNA sequence of the region of phage 2–22 (DS11S35) containing a TG microsatellite locus. The dinucleotide repeat is underlined and the sequences used for design of primers are boxed. (b) Autoradiograph of gel showing scoring of the D11S35 microsatellite in 18 unrelated individuals. Fragment sizes (nucleotides) are shown above the lanes. The four leftmost lanes contain a sequence ladder derived from a known DNA sequence.

be due to inhibition of template-independent extension by the nearby 5′-phosphate. However, in these experiments only one of the two primers was 5′-end-labelled and it is difficult to understand how this could inhibit template-independent addition to the distant 3′-end of this labelled strand.

The third mechanism, most evident when a single 5′-end-labelled primer is used, gives rise to a ladder of extra bands with a two-nucleotide spacing. The simplest way to explain this type of variability is by invoking slipped-strand mispairing during the PCR. Since slipped-strand mispairing should decrease at lower temperatures, one might expect that PCR of an end-labelled microsatellite using a DNA polymerase such as the Klenow enzyme, which is active at lower temperatures than the *Taq* polymerase, would yield fewer and less intense extra bands than those obtained with *Taq* polymerase.

The improvement in the quality of results that can sometimes be obtained when PCR products are synthesized using a single 5'-end-labelled primer is illustrated in *Figure 4*, which shows autoradiographs of PCR products from the DXS424 locus in Utah family K1333 obtained using internal labelling (*a*) and end-labelling (*b*).

An additional factor which can give rise to extra bands is the use of old labelled primers and/or PCR products. For best results, labelled primers should-be used within 2–3 days after kinasing and PCR products should be electrophoresed within 1–2 days after synthesis. (We have routinely stored PCR products at 4°C after adding dye-formamide mixture. It is possible that other storage conditions would improve the stability of these products.)

3. Potential improvements

3.1 Multiplexing

A major disadvantage of microsatellite VNTRs is that the scoring of each locus requires preparation of a sequencing gel. In contrast, scoring RFLPs (including 'classical' VNTRs) can be accomplished with reusable nylon membranes, allowing numerous loci to be scored on a single Southern blot. However, it should be possible to compensate for this to some extent by multiplexing. Multiplexing could be attempted at either or both of the two stages of the typing process (i.e. PCR amplification and resolution/ visualization of the allelic products). Multiplexing the first stage would be done by using multiple primer sets in single PCR reactions as has successfully been done by Chamberlain *et al.* (21; see also Chapter 5). The work of these authors indicates that sets of primer pairs would have to be carefully chosen and PCR conditions would have to be optimized to ensure that all of the multiplexed loci would be satisfactorily amplified.

Multiplexing the second stage would be done by loading mixtures of different PCR reactions in single gel lanes and blotting the resolved products on to nylon membranes. After UV crosslinking, the membranes would be probed sequentially with appropriate end-labelled PCR primers to reveal locus-specific fragments. A similar approach has already been used successfully for multiplex DNA sequencing (22). Because allelic fragments of microsatellite VNTRs tend to have rather narrow size ranges (19), it should be possible to use mixtures of several probes to visualize as many as three or four loci on a single autoradiogram. Since UV-crosslinked nylon membranes can be re-probed with oligomers approximately 50 times (22), it should be possible to type as many as 150 loci on a single membrane. Of course, the particular loci that could be typed would be predetermined by the PCR products originally loaded on the gel.

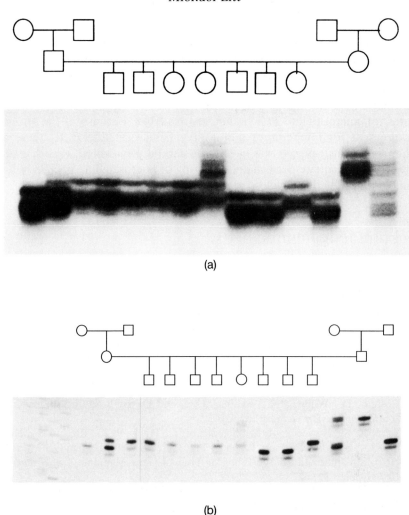

Figure 4. Autoradiograph of gels showing scoring of the DXS424 microsatellite in Utah family K1333 obtained using (a) internal labelling, and (b) end-labelling. *Figure 4B*, from reference 12, is reprinted with permission of the authors and the publisher.

3.2 Automation of the PCR

At present, scoring microsatellite VNTR loci in large numbers of individuals, such as the CEPH families, is a tedious, labour intensive task. If 96-well plate technology were available, much of the process could be automated. By using a robotic pipetting machine, such as a Biomek (Beckman Instruments), large numbers of PCR reaction mixtures could be prepared without manual pipetting or tube labelling. To make this practical, a thermal cycler capable of handling 96-well plates will be required. As of this writing, three such instruments are commercially available. These are (a) The BioOven from Biotherm Corp., (b) The PTC-100 from MJ Research, and (c) the Autogene, from Science/Electronics, or Grant Instruments (Cambridge) Ltd. Unfortunately, information on the actual performance of these instruments using 96-well plates for microsatellite PCR is not yet available. Because of the stringent requirement for temperature uniformity in microsatellite PCR, these and other similar instruments will have to be carefully evaluated. If such evaluation reveals adequate performance, scoring of microsatellite VNTRs in large numbers of individuals will be greatly facilitated.

Acknowledgements

This work was supported by a grant (GM 32500) from the National Institute of General Medical Sciences. V. Sharma contributed *Figure 1* and *3a* and J. A. Luty contributed *Figures 2, 3b*, and *4*. I thank S. Antonarakis, H. Dahl, Z. Guo, D. J. Kwiatkowski, X. Y. Luo, J. A. Luty, and B. Wieringa for contributing unpublished data included in *Table 1*.

References

1. Bostein, D., White, R. L., Skolnick, M., and Davis, R. W. (1980). *American Journal of Human Genetics*, **32**, 314.
2. Kidd, K. K., Bowcock, A. M., Schmidtke, J., Track, R. K., Ricciuti, F., Hutchings, G., Bale, A., Pearson, P., and Willard, H. F. (1989). Human Gene Mapping 10, Tenth International Workshop on Human Gene Mapping. *Cytogenetics and Cell Genetics* **51**, 622.
3. Jeffreys, A. J., Wilson, V., and Thein, S. L. (1985). *Nature*, **314**, 67.
4. Nakamura, Y., Leppert, M., O'Connell, P., Wolff, R., Holm, T., Culver, M., Martin, C., Fujimoto, E., Hoff, M., Kumlin, E., and White, R. (1987). *Science*, **235**, 1616.
5. Royale, N. J., Clarkson, R. E., Wong, Z., and Jeffreys, A. J. (1988). *Genomics*, **3**, 352.
6. Penrose, L. S. (1953). *Annals of Eugenics*, **6**, 33.
7. Weeks, D. E. and Lange, K. (1988). *American Journal of Human Genetics*, **42**, 327.
8. Litt, M. and Luty, J. (1989). *American Journal of Human Genetics*, **44**, 397.
9. Weber, J. L. and May, P. E. (1989). *American Journal of Human Genetics*, **44**, 388.
10. Litt, M., Sharma, V., and Luty, J. (1989). Human Gene Mapping 10, Tenth International Workshop of Human Gene Mapping. *Cytogenetics and Cell Genetics*, **51**, 1034.
11. Hofker, M. H., Skraastad, M. I., Bergen, A. A. B., Wapenaar, M. C., Bakker,

E., Millington-Ward, A., van Ommen, G. J. B., and Pearson, P. L. (1986). *American Journal of Human Genetics*, **39,** 438.

12. Luty, J. A., Guo, Z., Willard, H. F., Ledbetter, D. H., Ledbetter, S., and Litt, M. (1990). *American Journal of Human Genetics*, **46,** 776.
13. Tautz, D. (1989). *Nucleic Acids Research*, **17,** 6463.
14. Yandell, D. W. and Dryja, T. P. (1989). *American Journal of Human Genetics*, **45,** 547.
15. Miesfeld, R., Krystal, M., and Arnheim, N. (1981). *Nucleic Acids Research*, **9,** 5931.
16. Sambrook, J., Fritsch, E. F., and Maniatis, T. (1989). *Molecular Cloning, a Laboratory Manual*, 2nd edn. Cold Spring Harbor Laboratory Press, Cold Spring Harbor, NY.
17. Saiki, R., (1988). In *Genome analysis: A practical approach* (ed. K. E. Davies). IRL Press at Oxford University Press, Oxford.
18. Schmid, C. W. and Jelinek, W. R. (1982). *Science*, **216,** 1065.
19. Weber, J. L. (1989). In *The polymerase chain reaction*, Current Comments in Molecular Biology, Cold Spring Harbor Laboratory Press, Cold Spring Harbor, NY.
20. Oste, C. (1989). In *PCR technology* (ed. H. A. Ehrlich). Stockton Press, New York.
21. Chamberlain, S., Gibbs, R. A., Rainier, J. E., Nguyen, P. N., and Caskey, C. T. (1988). *Nucleic Acids Research*, **16,** 11141.
22. Church, G. M. and Kieffer-Higgins, S. (1988). *Science*, **240,** 185.
23. Church, G. M. and Gilbert, W. (1984). *Proceedings of the National Academy of Sciences, USA*, **81,** 1991.
24. Grunstein, M. and Hogness, D. (1975). *Proceedings of the National Academy of Sciences, USA*, **72,** 3961.
25. Benton, W. D. and Davis, R. W. (1977). *Science*, **196,** 180.
26. Thein, S. L. and Wallace, B. (1988). In *Human genetic diseases: A practical approach* (ed. K. E. Davies). IRL Press at Oxford University Press, Oxford.
27. Weber, J. L. and May, P. L. (1989). Human Gene Mapping 10, Tenth International Workshop on Human Gene Mapping. *Cytogenetics and Cell Genetics*, **51,** 1103.
28. Clark, J. M. (1988). *Nucleic Acids Research*, **16,** 9677.
29. Dean, M., Drumm, M. L., Stewart, C., Gerrard, B., Perry, A., Hidaka, N., Cole, J. L., Collins, F. S., and Ianuzzi, M. C. (1990). *Nucleic Acids Research*, **18,** 345.
30. Fujita, R., Hanauer, A., Sirugo, G., Heilig, R., and Mandel, J.-L. (1990). *Proceedings of the National Academy of Sciences, USA*, **87,** 1796.
31. Guo, Z., Sharma, V., Patterson, D., and Litt, M. (1990). *Nucleic Acids Research*, **18,** 4967.

Appendix to Chapter 6

Mouse microsatellites

JOHN A. TODD, MARCIA A. McALEER,
CATHERINE HEARNE, TIMOTHY J. AITMAN,
RICHARD J. CORNALL, and JEREMY M. LOVE

1. Introduction

There are at least twice as many TG/AC repeats in the mouse genome as in the human (1). We have shown recently that there is considerable allelic variation between microsatellites of different inbred strains of mice (2). Between the two strains DBA/2J and C57BL/6J, which have been used to construct one of the largest and most informative recombinant inbred (RI) strain mapping panels (3), 24/41 (58%) microsatellites were variable in size. The high degree of microsatellite variability between DBA/2J and C57BL/6J implies that a high resolution genetic map of the mouse genome can be constructed using RI strains. The advantage of RI strains for linkage analysis over genetic mapping using interspecific crosses is that the supply of DNA is limitless from RI strains. A disadvantage of RI strains is that DNA sequence variation between laboratory inbred strains, as detected by restriction enzymes, is difficult to find. Microsatellite variation will help alleviate this problem.

The high levels of DNA sequence variation between the wild species *Mus spretus* and laboratory strains is highly advantageous (4). PCR-analysed microsatellites offer additional variation which is easier to type than restriction fragment length variants in interspecific backcross genetic mapping as well as PCR typing requiring much less DNA. Twenty-four of 34 (70%) microsatellites were variable between *Mus spretus* and C57BL/6J (2). If 2.5 mg of DNA can be obtained from each interspecific backcross animal, then, at 250 ng of genomic DNA per PCR, 10000 markers could be analysed—well beyond what is required to achieve a 1-cM map of the mouse genome.

The localization of genes and analysis of their function are greatly facilitated by the use of the mouse as a manipulative experimental system. Large regions of the human and mouse genomes are conserved in the content and order of genes (5). Comparative analysis between mouse and human genomes is therefore a powerful approach to understanding gene function, particularly in disease.

We have chosen to use a non-radioactive method of detection of size variation between microsatellites after gel electrophoresis. About 50% of the first 50 microsatellites we have analysed are resolvable by agarose gel electrophoresis (2). The other alleles require resolution by acrylamide gel electrophoresis. Many variants in human and mouse microsatellites differ only by a few dinucleotide repeats. Unfortunately, acrylamide gel electrophoresis also resolves extra bands produced by the DNA polymerase which are often 2 bp or 4 bp smaller than the major product. These extra bands can complicate the scoring of alleles.

During searches of the nucleic acid databases we have found more complex microsatellites consisting of 3, 4, and 5 nucleotide repeat arrays (ref. 2, and unpublished). These are also variable between mouse strains and human individuals (S. Ghosh and J. A. Todd, unpublished data) and have the advantage over dinucleotide microsatellites in that size variation between alleles is greater and most alleles can be separated by agarose mini-gel electrophoresis.

2. Generation and detection of microsatellites

The methods used in our laboratory to study microsatellite VNTRs overlap considerably with those described by M. Litt in *Protocols 1–3* for the analysis of human microsatellites so only important differences are given in Section 2.1. Most of the mouse microsatellites we have analysed have been identified in the public databases (2) although 44 microsatellites randomly cloned from total mouse genomic DNA have also been characterized (unpublished data).

The use of ethidium bromide-staining or silver-staining to detect PCR products in non-denaturing gels avoids the use of radioactivity, which is a consideration when hundreds of reactions are run weekly. Gels consist of 6–10% acrylamide and are 0.75 mm thick and 16–20 cm long, so that they are convenient to handle during staining procedures. In addition, the results are obtained the same day and the method avoids some of the possible problems introduced by strand separation and radiolabelling discussed by M. Litt.

2.1 Analysis and detection of microsatellites

(a) Check the specificity of the PCR by a combination of titration of Mg^{2+} concentration and annealing temperature. Five 25 μl PCRs containing 1, 2, 3, 4, and 5 mM $MgCl_2$ are run for 32 cycles at [1 min at 94°C, 1 min at 55°C and 30 sec at 72°C], and a final 10-min extension at 72°C in a Perkin-Elmer/Cetus Cycler or LEP Prem III machine. The reaction mix contains 0.1% Tween-20 (Bio-Rad) and does not contain gelatin. Analyse PCR products by 4% agarose gel electrophoresis (3% NuSieve agarose (FMC)/1% agarose). Most primer pairs give a single band on agarose gel electrophoresis at a certain $MgCl_2$ concentration. If this is not the case, repeat the experiment at an annealing temperature of 60°C (and higher if necessary). If a satisfactory product is not obtained after these experiments then reduce primer concentrations, reduce the number of cycles, or finally design a third primer. Of over 100 mouse microsatellites analysed, fewer than 10% of microsatellites have been rejected (unpublished).

(b) Test the microsatellite for allelic variation by analysis of genomic DNA from different mouse strains at the established optimal Mg^{2+} concentration and annealing temperature using first, 4% agarose gel electrophoresis and, if no length variation is observed, non-denaturing acrylamide gel electrophoresis. Examples of three microsatellite variants that can be resolved by agarose gel electrophoresis are shown in *Figure 1*; (a) The *Ly-3* microsatellite on mouse chromosome 6 is a TG/AC repeat, (b) *Mpo* microsatellite on chromosome 11 is a trinucleotide repeat (GGA), and (c) the *Ckmm* microsatellite on chromosome 7 is a pentanucleotide (ATTTT) repeat.

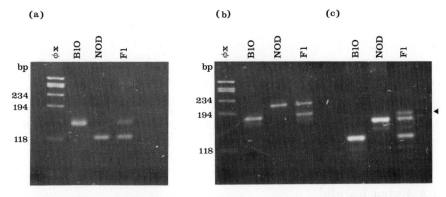

Figure 1. Ethidium bromide-stained 4% agarose gels showing size variants of the PCR-amplified microsatellites of the mouse loci (a) *Ly-3*, (b) *Mpo*, and (c) *Ckmm* using DNA from B10, NOD, and (NODXB10)F1 under the following Mg^{2+} concentrations and annealing temperatures, respectively: 2 mM, 55°C, 1 mM, 55°C and 2 mM, 55°C. Standard DNA markers are φx174 digested with *Hae*III (Gibco-BRL). The arrow in (c) denotes an additional higher molecular weight band which is specific to the F1 heterozygote and which is presumed to be a heteroduplex between strands from the two alleles.

(c) Load from 0.1 to 4 μl of the PCR in 1 × TBE/Bromophenol Blue/xylene cyanol/sucrose (5) on to a vertical, 6–10% acrylamide (19:1 acrylamide: bisacrylamide ratio)/1 × TBE gel. The gel is cast in a Bio-Rad Protean II apparatus with 16 or 20 cm-long plates with 0.75-mm spacers and 25 well-combs. Four gels can be run at the same time in a Protean II appparatus and, even at acrylamide concentrations of 6% are convenient to handle during the staining procedure. Size variation of 2 bp or greater can be resolved on 10 cm or longer acrylamide gels (2). The running buffer is 1 × TBE in the upper chamber and 0.5 × TBE in the lower chamber.

(d) Electrophorese PCR products at constant 80–350 V for 4–20 h with water cooling and buffer circulation using a magnetic stirrer.

(e) Visualize PCR products by either ethidium bromide-staining or, if the amount of DNA is low, silver-staining using the kit and instructions supplied by Bio-Rad. To reduce background staining with silver, do not use plasticware for any part of the silver-staining protocol and use filtered, deionized water. Handle gels with gloves that have been washed in filtered, deionized water. An example of a mouse TG/AC microsatellite, called *Plau* (2) on chromosome 14, resolved by acrylamide gel electrophoresis and detected by silver-staining, is shown in *Figure 2*.

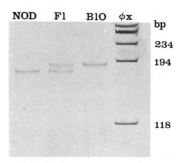

Figure 2. Silver-stained 8% acrylamide gel showing the size variant of the PCR-amplified microsatellite of the mouse locus *Plau* using DNA from NOD, B10, and (NODXB10)F1 under the following PCR conditions: 2 mM MgCl$_2$, 55°C annealing. Standard DNA markers are φx174 digested with *Hae*III (Gibco-BRL).

2.2 Extra bands

After acrylamide gel electrophoresis extra bands around the major PCR product are often observed. The only way we have found to reduce the intensity of these extra bands is to titrate the number of PCR cycles. If a primer pair at 32 cycles of the PCR produces extra bands which prevent unambiguous scoring then the PCR is run with 25 cycles. The extra bands usually only appear when the DNA is subject to too many PCR cycles. Reduction of primer concentration (to about 0.2 μM) can also improve the quality of the PCR product.

2.3 Instrumentation

We use two different temperature cycling machines: a Perkin-Elmer/Cetus Thermal Cycler and a LEP Prem III (Anderman Ltd.). The Prem III has the advantage of having two types of block: one for 0.5-ml microfuge tubes and the other for 96-well plates (Falcon 3913). The blocks are interchangeable and we have observed no differences in PCR efficiency between tubes and plates. PCRs are set up in plates in a similar manner to tube reactions: water first, DNA next followed by the PCR mix with the other reaction components. Furthermore, the Prem III uses a heating element to heat the block and a Peltier device to cool the block, which may be an advantage over other machines that use Peltier devices to heat and cool where longevity of the device is questionable.

2.4 Future trends

VNTRs that can be resolved by agarose gel electrophoresis are much easier to use than those that have to be resolved by acrylamide gel electrophoresis (see *Figure 1*). Other examples of variable mouse microsatellites that contain repeat motifs greater than 2 bp include a (CAAG) repeat in the *Cyp1a2* locus on chromosome 9 (3) and a (GGGAGA) repeat in the *Igh* locus on chromosome 12 (unpublished data). These larger repeat motifs offer a possible solution to problems associated with the resolution of length differences less than 10 bp that are often complicated by extra bands generated by the DNA polymerase.

Acknowledgements

This work was supported by the Medical Research Council (UK) as part of the UK Human Genome Mapping Project, the Wellcome Trust and the Juvenile Diabetes Foundation. Discussion with our colleagues including A. Jeffreys is gratefully acknowledged.

References

1. Hamada, H., Petrino, M. G., and Kakunaga, T. (1982). *Proceedings of the National Academy of Sciences, USA*, **79**, 6465.
2. Love, J. M., Knight, A. M., McAleer, M. A., and Todd, J. A. (1990). *Nucleic Acids Research*, **18**, 4123.
3. Taylor, B. A. (1989). In *Genetic variants and strains of the laboratory mouse* (ed. M. F. Lyon and A. G. Searle), pp. 773–96. Oxford University Press, Oxford.
4. Avner, P., Amar, L., Dandolo, L., and Guenet, J.-L. (1988). *Trends in Genetics*, **4**, 18.
5. Sambrook, J., Fritsch, E. F., and Marriatis, T. (1989) *Molecular cloning: a laboratory manual*, (2nd edn.). Cold Spring Harbor Laboratory Press, NY.

7

Genome amplification using primers directed to interspersed repetitive sequences (IRS-PCR)

SUSAN A. LEDBETTER and DAVID L. NELSON

1. Introduction

The polymerase chain reaction (PCR) has revolutionized the isolation and analysis of specific nucleic acid fragments from a wide variety of sources (1). However, application of the PCR to isolate and analyse a particular DNA region has required knowledge of DNA sequences flanking the region of interest. This limits amplification to regions of known DNA sequence. We sought to amplify human DNA of unknown sequence specifically from somatic cell hybrids retaining human chromosome fragments in rodent cell backgrounds. This allows the isolation and characterization of sequences from specific human regions retained in hybrids, obviating the requirement for cloned DNA libraries and isolation of human clones through the use of human-specific repeat sequence probes (2). It also allows the characterization of hybrid cell lines for quantity and quality of the human sequences retained. The method, interspersed repetitive sequence PCR (IRS-PCR; 3–5), has also proved useful for the rapid isolation of human DNA inserts from cloned sources, extending the application of the PCR to genomic DNAs cloned in lambda and yeast artificial chromosome (YAC) (6) vectors. The adaptation of the PCR to large genomic regions provides another tool for the analysis of the human genome.

The IRS-PCR method makes use of the ubiquitous *Alu* and L1Hs repeat sequences found in human DNA. Approximately 900 000 copies of the 300 bp *Alu* sequence are distributed throughout the human genome (7). Although there is considerable variation between copies of the *Alu* repeat, a consensus sequence has been established, and there are regions of the repeat that are reasonably well conserved (8). L1Hs (also referred to as *Kpn*I) is the major human LINE repeat element, found in 10 000–100 000 copies of a complete or truncated 6.4 kb sequence (9).

Rodent genomes contain repetitive sequences similar to *Alu* and L1Hs repeats. In order to amplify human DNA specifically from somatic cell hybrids in rodent cell backgrounds, it is necessary to identify regions of these repetitive sequences that are sufficiently divergent between human and rodent repeats. Another consideration is the conservation of primer binding sites among copies of the repeat sequence in the human genome. Thus, primers were chosen for a high level of conservation in human repeats and as little homology as possible between human and rodent version of the repeat. In the *Alu* repeat sequence, a 31 bp insert in the second monomer of the dimeric human repeat is not found in rodent monomeric *Alu* equivalents (10). The 5′ 17 bp of this 31 bp region is well-conserved among human repeats, and these 17 bp have been exploited for design of primers *Alu*-559 and *Alu*-517. The L1Hs repeat is highly variable at the 5′-end, however the 3′-end is relatively constant, and a 208 bp region has been described with high homology between human copies of the repeat and with no cross-species homology (11). This region was chosen for the L1Hs primer sequence.

2. Choice of primer sequences

Three different human-specific primer sequences are presented here. The two *Alu* primers are complementary and allow amplification in either orientation from the *Alu* repeat. It is important to stress that *Alu*-517 and *Alu*-559 cannot be used in the same reaction since they are complementary and will merely amplify a 'primer dimer' (12). Thus, the reactions that utilize a single *Alu* primer can amplify only those sequences found between adjacent copies of *Alu* repeats in inverted orientation (see *Figure 1*). Each *Alu* primer can be used in conjunction with the L1Hs primer (4) to generate products between L1Hs and *Alu* repeats.

> L1Hs   CATGGCACATGTATACATATGTAAC(T/A)AACC
>
> *Alu*-517   C̲G̲A̲C̲C̲T̲C̲G̲A̲G̲ATCT(C/T)(G/A)GCTCACTGCAA
>
> *Alu*-559   A̲A̲G̲T̲C̲G̲C̲G̲G̲C̲C̲G̲CTTGCAGTGAGCCGAGAT

Both *Alu* primers include 5′ extensions (underlined) that are unrelated to the *Alu* repeat. These include restriction sites for cloning the products. *Alu*-517 includes *Xho*I and *Bgl*II cloning sites; *Alu*-559 includes a *Not*I cloning site. Primers without these 5′ extensions are not human specific for reasons that are unclear to us. Other 5′ extensions (with other restrictions sites for instance) also allow these primers to amplify in a human specific manner. Thus, the presence of a 5′ extension (increasing the total length to a minimum of 23 bp) rather than the specific sequence is important for the human specificity.

Figure 2 shows examples of amplification using the three primers described to generate PCR products from two different somatic cell hybrid DNAs with either an intact (Nl 17) or fragmented (PLT6B; 13) human chromosome 17 in a mouse cell background. Amplification with either *Alu*-559 or *Alu*-517

Susan A. Ledbetter and David L. Nelson

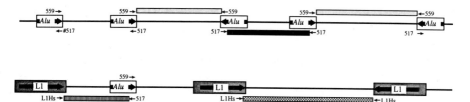

Figure 1. IRS-PCR. Amplification with repeat sequence primers proceeds between adjacent copies of repeat elements oriented in the appropriate direction. Shown are the three primers used for human specific amplification from somatic cell hybrids (*Alu*-517, *Alu*-559, and L1Hs).

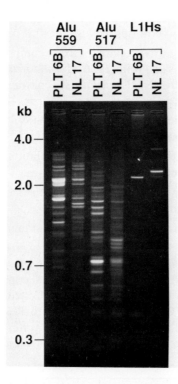

Figure 2. IRS-PCR of two hybrids retaining human chromosome 17 using two *Alu* and one L1 primer. Hybrid cell lines retaining human chromosome 17 were amplified with three different primers in single primer IRS-PCR and subjected to agarose gel electrophoresis as described in Section 4.1. Primers used are listed above each lane. *Alu*-559 and *Alu*-517 are human specific *Alu* primers, while L1Hs is a human specific primer derived from 3' consensus sequences of the L1Hs repeat family. The L1HS primer shows much less amplification when used alone due to the lower frequency of L1Hs repeats. The hybrid PLT 6B contains portions of 17p and proximal 17q (13), while Nl 17 is a hybrid with an intact chromosome 17. Both are retained in mouse cell backgrounds.

109

as single primers generates considerably more product than L1Hs used alone. This is a reflection of the difference in abundance of the *Alu* and L1Hs repeat sequences; it is rare to find two copies of the L1Hs repeat in inverted orientation and within an amplifiable distance. Use of the L1Hs primer in conjunction with either *Alu* primer generates a different pattern of products that is equal or greater in complexity to that seen with either *Alu* primer used alone (not shown).

3. PCR conditions

3.1 PCR amplifications

All reaction conditions have been optimized for the Perkin-Elmer/Cetus DNA Thermal Cycler:

(a) Sterile tubes and pipette tips are used for all manipulations.

(b) It is most convenient to prepare a mix of components to distribute to all tubes receiving the same primers. Add components in the following order: Water, 10 × buffer, dNTPs, primer(s), enzyme. Mix well by vortexing.

(c) The mix is then distributed to reaction tubes (95–100 μl per tube, dependent upon the concentrations of the DNA. (Variations of a few microlitres in the amount of DNA added do not have an effect.)

(d) DNA is added to each tube. Leaving manipulation of DNA until the last step minimizes cross-contamination.

(e) Mineral oil (80–100 μl) is added (be sure to change pipette tips between additions to reaction tubes).

(f) Tubes are placed in the Thermal Cycler programmed with the conditions in *Protocol 1*.

Amplification of somatic cell hybrid DNAs are performed according to *Protocol 1*. L1Hs and *Alu*-559 are used at a final concentration of 1 μM. *Alu*-517 is used at a final concentration of 0.3 μM. Concentrations for primers are calculated based on an OD_{260} of 1 being equal to 30 μg/ml. (1 base equals 330 daltons.) Primers are resuspended in distilled water after de-protection and lyophilization. We have found no advantage in the use of gel purified oligonucleotides.

Protocol 1. Amplification reaction

1. Mix in a 0.4 ml microcentrifuge tube:

- ~1 μg DNA
- 10 μl 10 × PCR buffer[a]
- 1 μl 25 mM dNTPs[b] (final concentration of 250 μM each dNTP)

- 1 µl 100 × primer (see concentrations above)
- 0.5 µ AmpliTaq polymerase (Perkin-Elmer/Cetus)
- Water to 100 µl

2. Overlay with ~80 µl light mineral oil.

3. Use the following reaction conditions:

95°C, 4 min (Initial denaturation)
(94°C, 1 min, 55°C, 1 min, 72°C, 4 min × 30–35 cycles)
72°C, 7 min (Final extension)

[a] 10 × PCR buffer: 100 mM Tris–HCl, pH 8.4, 500 mM KCl, 15 mM $MgCl_2$, 0.01% w/v gelatin: this is the buffer recommended by Cetus.
[b] 25 mM dNTPs solution is prepared with an equal volume of a 100 mM stock of each of dATP, dCTP, dGTP, and dTTP—we use Pharmacia nucleotides purchased in solution at 100 mM.

3.2 Consideration of primer and dNTP concentration

The primer concentrations are based on amplification of 100–1000 ng of hybrid DNA. When using significantly less DNA it may be necessary to reduce the primer concentration to avoid amplifying rodent sequences.

A surprising amount of variation between different batches of 10 × buffer has been found even when using the same stock solutions for preparation. The variability ranges from excellent to no amplification. It is recommended that at least 10 ml of buffer be made at a time. Each lot can be tested and good ones stored in 1 ml aliquots at −20°C for months.

Each time new buffer is made it is useful to titrate the dNTP concentration by amplifying a known hybrid (we use one containing the X chromosome as its only human material) and the rodent parent using final dNTP concentrations of 125, 250, and 300 µM. This allows comparison of the PCR pattern with the current buffer. Ideally, maximum amplification of the human material with no amplification of the rodent parent is found. If none of these concentrations give human specific amplification, a lower primer concentration is attempted. It should be possible to achieve human specific amplification with these primers at a 55°C annealing temperature by varying the dNTP concentration and/or the primer concentration.

4. Analysis of IRS-PCR products

4.1 Gel analysis of PCR products

Maximum resolution of gel analysis of the complex patterns of hybrid amplification requires some post-PCR manipulation (*Protocol 2*). In particular, Klenow extension and precipitation of the PCR gives a much clearer gel with less background smear and sharper bands. The Klenow treatment allows completion of unfinished extension products and the precipitation eliminates most of the unincorporated dNTPs and primer. Both steps reduce the complex background seen in untreated products. This becomes particularly important when trying to determine if a faint fragment is present or absent.

Protocol 2. Post PCR treatment of samples

1. Transfer approximately 90 µl of the PCR product to a clean microcentrifuge tube leaving the mineral oil behind.

2. Incubate the PCR product with ~2 units of Klenow fragment for 20 min at 37°C.

3. Following the 37°C incubation place the tube at 65°C for 10 min to inactivate the enzyme.

4. Precipitate the PCR product by adding an equal vol. of 2M ammonium acetate and 3 vol. of absolute ethanol. Incubate at room temperature for 5 min then centrifuge for 10 min in a microcentrifuge.

5. Resuspend the pellet in about 80 µl of TE (10mM Tris–HCl, pH8.0, 1mM Na_2 EDTA).

6. Run 5–10 µl of the PCR product on an agarose gel containing 0.5 µg/ml ethidium bromide and 1 × TBE.[a] Use a 1.3% agarose gel for any of the three primers used alone, and a 3% NuSieve (FMC Bioproducts) gel for reactions using L1Hs and *Alu* primers together as fragments tend to be smaller.

7. Photograph the gel on Polaroid type 55 film or equivalent (positive/negative) by a 30–45 sec exposure on a short-wave UV transilluminator. Type 57 film does not provide adequate resolution for detailed analysis although it can be used to decide that the PCR has worked before the Klenow step and precipitation described in this *Protocol*.

[a] 10× TBE is 90mM Tris, 89mM boric acid, (pH8.3), 2.5mM EDTA. Resolution of the products will not be as good in TAE buffer and is very poor if the gel is run without EtBr.

4.1.1 Interpretation

The IRS-PCR pattern of a given chromosome appears to be fairly constant. We have compared five unrelated X chromosomes as well as two each of chromosomes 7, 17, 19, and Y, and found little or no variation. There can be variation in intensity of a given band when comparing different amplifications of the same chromosome but it should always be present. It is best to include all controls and unknowns in the same PCR experiment since the variations seem to be between different runs rather than in the same run. The use of a master mix of all the reaction components except the DNA ensures consistency in the amplification reactions; thus differences between samples are due to the DNA added.

An example of products that can be unambiguously assigned by the presence in IRS-PCR of hybrids differing in the amount of a chromosome retained is seen in *Figure 3*. This shows amplification reactions using the *Alu*-517 primer of a series of hybrid cell lines retaining portions of human chromosome 11p in a hamster cell background (14). The region indicated by a bracket contains a number of PCR products, all present in the parental hybrid (*lane 1*), and progressively missing from hybrids with smaller regions of

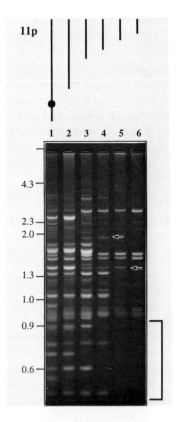

Figure 3. Amplification of hybrids retaining fragments of human chromosome 11p. Six hybrid cell lines retaining portions of human chromosome 11 (depicted above each lane) in a hamster cell background (14) were amplified with *Alu*-517 and subjected to agarose gel electrophoresis as described in Section 4.1. Arrows indicate amplification products that are visible only in hybrids with small portions of the chromosome. The upper arrow marks a fragment that is clearly visible in **lanes 3** and **4**, and is not visible in **lanes 1** and **2** due to the numerous bands generating a smear in this region. The lower arrow marks a fragment clearly visible in **lanes 3–5** that is covered by a more prominent product in **lanes 1** and **2**. The bracket delineates an area of the gel where a number of products, unique to regions of the chromosome that are successively lost in the hybrids, can be seen. **Lanes 1–6** contain hybrids J1–11, J1–14, J1–52a, J1–2, J1–1, and J1–43a respectively.

113

human chromosome 11p. The product at 600 bp, for example, is present in the hybrids in *lanes 1–3* and absent in those in *lanes 4–6*. This allows assignment of this product to the region of the chromosome between the breakpoints in the hybrids in *lanes 3* and *4*. Similarly, products mapping to intervals defined by the hybrids in *lanes 1* and *2*, *2* and *3*, and *4* and *5* can be seen in this region of the gel.

When trying to compare a hybrid containing a portion of a chromosome to its parent there are several points to be aware of; these are also illustrated in *Figure 3*. (a) When the amount of human DNA is reduced by deleting part of the chromosome the amplification of the remaining fragments should be at least as bright as in the whole chromosome and are usually much brighter; the fewer potential templates available for amplification the better any given template will amplify. (b) New fragments that were not apparent in the whole chromosome will now be seen; for example the product indicated by the upper arrow in *Figure 3*. This fragment becomes visible only after the reduction in complexity between the hybrids in *lanes 2* and *3*. (c) If a fragment in the reduced hybrid is fainter than in the whole chromosome it is likely that one of two comigrating fragments has been lost. This phenomenon is illustrated in *Figure 3* where the product indicated by the lower arrow is obscured in *lanes 1* and *2*, and is only clearly visible in hybrids (*lanes 3–5*) where the template accounting for the bright band seen in *lanes 1* and *2* is no longer present.

4.2 Analysis of PCR fragments by hybridization in complex hybrids

Protocol 2 is designed to identify fragments in specific chromosomal regions by analysing the ethidium-stained PCR products directly on a gel. This generally requires that all of the hybrids used in the analysis are monochromosomal since the addition of other templates to the reaction alters the pattern of amplification. Most regional mapping panels for a given chromosome contain translocations or deletions of the chromosome of interest; however, they frequently also contain additional chromosomes. The requirement for single chromosome hybrids for the PCR analysis can be obviated by Southern analysis of the PCR products from the mapping panel hybrids using labelled products from the hybrid containing the region of interest (*Protocol 3*).

Figure 4 shows the result of hybridizing the ^{32}P-labelled *Alu*-517 PCR product of a hybrid (PLT6B) that contains portions of 17p and proximal 17q to a filter which contains a hybrid that has the intact 17 as its only human material (Nl 17) and to four hybrids that contain deletions of 17p13. These four hybrids all contain 8–14 additional human chromosomes. The iso17q hybrid has 2–3 additional human chromosomes. The fragment at 1.3 kb that appeared to be missing in the del17p13 and MDS-5 hybrids was isolated from a low-melting-temperature gel of the PCR products of PLT6B and used as probe on a Southern blot of *Eco*RI digested DNA of the hybrids which confirmed its location in 17p13.3 (not shown).

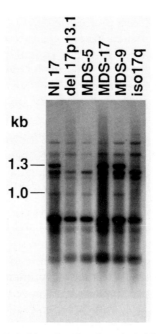

Figure 4. Analysis of complex hybrids using Southern hybridization of PCR product with a PCR product probe. A probe derived from *Alu*-517 amplified PLT6B (see *Figure 2*) was used to hybridize against *Alu*-517 amplified products from a number of hybrid cell lines retaining chromosome 17 sequences using conditions described in *Protocol 3*. As many of the hybrids analysed retain multiple human chromosomes in addition to 17, hybridization allows regional assignment of chromosome 17-specific products that is not possible by inspection of the ethidium bromide-stained gel alone. The 1.3 kb product missing in del 17p13.1 and MDS-5 was confirmed to be derived from 17p13.1 by Southern hybridization to a mapping panel of restricted DNAs using the isolated fragment from the PLT6B amplification reaction. NI17 contains an intact chromosome 17 as the only human material, del 17p13.1, MDS-5, MDS-17, and MDS-9 all retain chromosomes 17 with deletions of 17p13 as well as 8 to 14 additional human chromosomes, and iso17q also contains 2 to 3 other human chromosomes.

It is possible to use the PCR products of an intact chromosome as the probe, but the background is generally high and faint fragments will be obscured. The best hybrid to use for generating a probe is one that contains the region of interest and as little material from the rest of the chromosome as possible.

Protocol 3. Hybridization analysis

1. Perform the PCR to be used for hybridization analysis as described in *Protocol 1* then treat the products with Klenow and precipitate according to *Protocol 2* (steps 1 to 4).

2. Resuspend the pellet from 90 μl of the original reaction in 60 μl TE for

Protocol 3. *Continued*

monochromosomal hybrids, 30 μl for hybrids containing up to 5 additional chromosomes and 15 μl for hybrids containing >5 chromosomes. Run 5 μl of each on a 1.3% agarose gel using a 1 mm comb. The gel should run until the dye front has migrated ~13 cm.

3. Photograph the gel using Kodak type 57 or equivalent film to minimize UV exposure.

4. Denature in two changes of 0.4 N NaOH for 20 min each (neutralization is not necessary) and then transfer to GeneScreen Plus (DuPont) in 10 × SSC.

5. When the transfer is complete, place the filter in 0.4 N NaOH for ~1 min followed by 0.2 M Tris–HCl, pH 7.4, 2 × SSC for 5 min.

6. Allow the filter to dry briefly and then pre-hybridize for at least 4 h in 1 M NaCl, 1% SDS, 10% dextran sulphate, and 0.5 mg/ml human placental DNA at 65°C.

7. Precipitate the PCR products of the hybrid to be used as probe (*Protocol 2*, step 4) then label by the random primer method (17) followed by spermine or ethanol precipitation to remove the unincorporated dNTPs.

8. To preassociate the probe (10^6 c.p.m./ml) mix with 2–3 ml of hybridization solution containing 0.5 mg/ml placental DNA, boil for 5 min then incubate at 65°C for 2 h.

9. Add the probe to the filter and hybridize at 65°C overnight.

10. Wash the filter for 30–45 min each in 2 × SSC/0.1% SDS, 1 × SSC/0.1% SDS and 0.1 × SSC/0.1% SDS, at 65°C.

11. Expose the filter to Kodak X-AR or equivalent film with an intensifying screen for ~1 h.

12. To reprobe the filter strip bound probe by boiling in 0.1% SDS/0.1 × SSC for 30 min. This step can be repeated if all of the probe is not removed.

This method for detection and isolation of regionally localized PCR products is quite sensitive once the conditions have been optimized. Even in the best circumstance, however, a decision on a given fragment being present or absent may be wrong. Using this method to identify probes in 17p13.3 (by gel analysis and hybridization) has allowed 10 to 18 fragments initially thought to be missing in 17p13 to be confirmed. The remaining 8 fragments mapped to other regions of 17. The overall yield compared to the small amount of work required to isolate the clones was very good. We have also used this method to characterize a series of hybrids that are deleted for various amounts of the X chromosome and have been able to distinguish between lines that had previously been found to differ by the presence or absence of a single probe by Southern analysis.

5. Preparing cells for PCR

The PCR can be performed on lysed cells (*Protocol 4*) as well as purified DNA. We have found it convenient for early analysis of hybrid cell lines to perform IRS-PCR on the hybrid cell colonies early in their propagation to determine the amount of human sequence present (if any). This analysis may also provide an early opportunity to characterize the hybrids. We are currently using cell lysates only to determine if human DNA is present in a new hybrid and so we have not yet optimized this procedure for detailed analysis.

Protocol 4. Preparation of cells for PCR

1. Spin down $\sim 10^5$ cells.[a]
2. Wash once with $1 \times$ PCR buffer—remove as much of wash as possible.
3. Store pellet at $-20°C$ or process immediately.
4. Resuspend pellet in 20 µl of lysis buffer[b] plus 1 µl of 1 mg/ml proteinase K.
5. Incubate at $50–60°C$ for 1 h.
6. Place at $95°C$ for 10 min to inactivate proteinase K.
7. Spin briefly to pellet cell debris.
8. Use 5 µl of supernatant for each 100 µl PCR reaction (*Protocol 1*).
9. Store supernatants at $-20°C$.

[a] The procedure can be scaled up or down using 20 µl lysis buffer per 10^5 cells as a guide.
[b] Lysis buffer: 1 ml $10 \times$ PCR buffer, 45 µl NP-40 (Nonidet P-40, Sigma), 45 µl Tween-20, 8.9 ml sterile H_2O. Store at $4°C$.

---

6. IRS-PCR applied to cloned DNA

The *Alu* and L1Hs primers have also been used to generate fragments from cloned human DNA in *E. coli* vectors (3) and yeast (15) cloning systems. The reaction conditions for *Alu-Alu* or L1-*Alu* amplification are identical to those for somatic cell hybrids. Since these reactions begin from cloned DNA, much less DNA is required. For clones in cosmid or bacteriophage vectors, amplification can begin with colonies or plaques (16). For clones in yeast artificial chromosome vectors, it is best to use purified DNA from the yeast cell harbouring the YAC, but 10–50 ng of DNA is sufficient. Successful amplification has been observed from yeast colonies; however, these reactions have been quite variable.

For amplifying human insert sequences from cloned fragments, it is often valuable to include primers that recognize vector sequences immediately adjacent to the inserted DNA. This allows amplification from the vector to a

117

repetitive sequence some distance from the cloning site. In this way it is possible to generate specific end fragments from cloned DNA which are useful for detecting overlapping clones (chromosome walking). Below are listed several primers that have been used successfully for the generation of end-specific fragments from bacteriophage, cosmid and YAC vectors. In these reactions, one vector primer is typically employed with two *Alu* primers, allowing amplification in both orientations from *Alu* repeats in the insert. The *Alu*-278 primer listed below is employed for amplification in place of *Alu*-517, since it is close to the 5′ end of the *Alu* sequence, allowing amplification of less of the repeat than *Alu*-517. *Alu*-278 is not human-specific in somatic cell hybrids; however, in cloned sequences this is unimportant.

6.1 Primers for vector-*Alu* amplification.

- *Alu*-278: CCGAATTCGCCTCCCAAAGTGCTGGGATTACAG
  Alu 5′-end primer.
- T3: ATTAACCCTCACTAAAG
- T7: AATACGACTCACTATAG
  For cosmid and phage vectors with T3 and T7 RNA polymerase promotors (e.g. pWE (17), λDASH Stratagene). Since they are only 17 bp in length, amplifications including these primers are carried out at 50 °C.
- YAC4L: CGGAATTCGCCAAGTTGGTTTAAGGCGCAAGAC
- YAC4R: GGAAGCTTGGCGAGTCGAACGCCCGATCTCAAG
  For the YAC cloning vector pYAC$_4$ (15).

5′ extensions with restriction sites added for cloning PCR products are underlined.

Acknowledgements

We wish to thank C. Thomas Caskey and David H. Ledbetter for advice and support in the development of these methods, Maureen F. Victoria for critical reading of the manuscript, Carol Jones for the contribution of J1 cell DNAs and Ellen Solomon for the PLT6B hybrid. This work was supported by grants from the US Department of Energy (DE-FG05-88ER60692), and the National Institutes of Health (HD20619), and by the Howard Hughes Medical Institute.

References

1. Mullis, K. B. and Faloona, F. A. (1987). In *Methods in Enzymology* (ed. R. Wu). Vol. 155, pp. 335–50. Academic Press, New York and London.
2. Gusella, J. F., Keys, C., Varsanyi-Breiner, A., Kao, F.-T., Jones, C., Puck,

T. T., and Housman, D. (1980). *Proceedings of the National Academy of Sciences, USA*, **77**, 2829.

3. Nelson, D. L., Ledbetter, S. A., Corbo, L., Victoria, M. F., Ramirez-Solis, R., Webster, T. D., Ledbetter, D. H., and Caskey, C. T. (1989). *Proceedings of the National Academy of Sciences, USA*, **86**, 6686.
4. Ledbetter, S. A., Nelson, D. L., Warren, S. T., and Ledbetter, D. H. (1990). *Genomics*, **6**, 475.
5. Wilson, A. and Goodfellow, P. J. (1989). *American Journal of Human Genetics*, **45**, A168.
6. Burke, D. T., Carle, G. F., and Olson, M. V. (1987). *Science*, **236**, 806.
7. Britten, R. J., Baron, W. F., Stout, D. B., and Davidson, E. H. (1988). *Proceedings of the National Academy of Sciences, USA*, **85**, 4770.
8. Kariya, Y., Kato, K., Hayashizaki, Y., Himeno, S., Tarui, S., and Matsubara, K. (1987). *Gene*, **53**, 1.
9. Singer, M. F. and Skowronski, J. (1985). *Trends in Biochemical Sciences*, **10**, 119.
10. Jelinek, W. R. and Schmid, C. W. (1982). *Annual Reviews in Biochemistry*, **51**, 813.
11. Scott, A. F., Schmeckpeper, B. J., Abdelrazik, M., Comey, C. T., O'Hara, B., Rossiter, J. P., Cooley, T., Heath, P., Smith, K. D., and Margolet, L. (1987). *Genomics*, **1**, 113.
12. Saiki, R. K. (1989). In *PCR Technology* (ed. H. A. Erlich) pp. 7–16. Stockton Press, New York.
13. Xu, W., Gorman, P. A., Rider, S. H., Hedge, P. J., Moore, G., Prichard, C., Sheer, D., and Solomon, E. (1988). *Proceedings of the National Academy of Sciences, USA*, **85**, 8563.
14. Glaser, T., Houseman, D., Lewis, W. H., Gerhard, D., and Jones, C. (1989). *Somatic Cell and Molecular Genetics*, **15**, 477.
15. Nelson, D. L., Ballabio, A., Victoria, M. F., Pieretti, M., Bies, R. D., Gibbs, R. A., Maley, J. A., Chinault, A. C., Webster, T. D., and Caskey, C. T. (1991). Proceedings of the National Academy of Sciences, USA, (in press.)
16. Saiki, R. K., Gelfand, D. H., Stoffel, S., Scharf, S. J., Higuchi, R., Horn, G. T., Mullis, K. B., and Erlich, H. A. (1988). *Science*, **239**, 487.
17. Wahl, G. M., Lewis, K. A., Ruiz, J. C., Rothenberg, B., Zhao, J., and Evans, G. A. (1987). *Proceedings of the National Academy of Sciences, USA*, **84**, 2160.

PCR amplification of microdissected DNA

DANIEL H. JOHNSON

1. Introduction

A common experimental strategy for understanding the molecular basis of developmental and physiological processes begins with the isolation of the genes encoding the cellular components which mediate these events. In many cases our knowledge of these genes is limited to the relative physical map location, on a chromosome, of mutant alleles which appear to perturb normal processes or contribute to disease states. With the advent of *in vitro* amplification synthesis of DNA (PCR) it has been possible to develop efficient protocols for the recovery of DNA restriction fragments from microdissected chromosomal regions harbouring specific genes of interest (1, 2). Such probes offer a potent alternative to chromosome walking for the recovery of genomic DNA clones encompassing a defined region of the chromosome.

The goal of this chapter is to present the details of a strategy for amplification of DNA molecules whose sequence is unknown and thus could not be amplified using standard PCR protocols. The method is generally useful for producing an amplifiable collection of DNA fragments from any source of DNA whether it be a chromosome fragment or the entire genome of an organism. The chapter will concentrate on the use of the method to amplify microdissected DNA with a final section on whole genome amplification. A detailed presentation of all relevant procedures involved in microdissection or standard molecular biology protocols necessary for the practice of the method is beyond the scope of this chapter. These have been previously reported in great detail elsewhere and are referenced throughout the chapter.

2. General features of the method

The major steps of the procedure are outlined in *Figure 1*. The general strategy is to extract DNA from chromosome fragments, chromosomes or organisms that have been microdissected from cytological specimens fixed to a glass coverslip. Dissected material is placed in a few nanolitres of aqueous

121

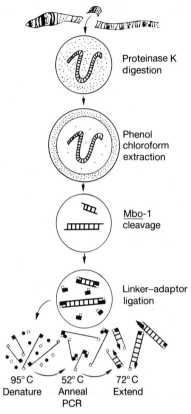

Figure 1. A cartoon outlining the method for amplification of DNA from dissected chromosome fragments. Microscopic steps of the procedure are performed in a paraffin oil-filled chamber to prevent evaporation. In the PCR step, large black boxes represent the 20-mer primer binding site. Small black boxes represent the primer and black circles represent the *Taq* polymerase.

buffer held in a paraffin oil-filled chamber to maintain an effective concentration of the DNA and other reagents while preventing evaporation of the sample during the micro-volume steps of the procedure. The DNA is digested to completion with one or more restriction endonuclease and a linker-adapter is ligated at both ends. The linker provides an identical primer binding site on each strand of the target DNA permitting subsequent amplification of the linker-modified restriction fragments using PCR (3).

2.1 Selection of restriction endonucleases

The choice of restriction endonucleases to be used in this method is dictated primarily by the template length preferences shown in PCR for complex mixtures of different sized templates. In practice it is possible to amplify molecules as large as several kilobase pairs, however large templates are strongly discriminated against when amplification occurs in a mixture including shorter, more efficiently amplified templates. This is particularly a problem when the starting number of target molecules is very low, necessitating 40 or more cycles of amplification to produce a useful mass of DNA from microdissected material. Since the goal in most cases is to recover as much of the dissected DNA sequence as possible in the amplified product, it is desirable to restrict the DNA with one or more enzymes to produce fragments of 500 bp in average length. This is most easily accomplished by using restriction endonucleases with four base pair recognition sequences and which leave sticky ends for efficient ligation of the linker-adapter (*Table 1*). A simple strategy to further augment the sequence complexity present in the amplified DNA is to digest multiple dissected aliquots of the target with different restriction endonucleases. This approach increases the likelihood of including any specific DNA sequence present in the dissected target as part of an efficiently amplified PCR template in at least one of the restriction digests. Note that the isoschizomer pairs *Sau*3A/*Mbo*I and *Msp*I/*Hpa*II (*Table 1*) have different restriction site methylation requirements for cleavage and thus produce substantially different spectrums of restriction fragments from the same complex eucaryotic DNA target.

Table 1

Restriction enzyme	Cleavage consensus	Linker-adapter
*Mbo*I[a]	5′− ↓ GATC −3′	5′GATCTGTACTGCACCAGCAAATCC3′
*Sau*3A[b]	3′− CTAG ↑ −5′	3′ ACATGACGTGGTCGTTTAGG5′
*Hpa*II[c]	5′−C ↓ CG G−3′	
*Msp*I[d]	3′−G GC ↑ C−5′	5′CGGTGTACTGCACCAGCAAATCC3′
		3′ CACATGACGTGGTCGTTTAGG5′
*Taq*I[e]	5′−T ↓ CG A−3′	
	3′−A GC ↑ T−5′	

[a] No cleavage with N^6-methyladenine but cleaves with 5-methylcytosine.
[b] No cleavage with 5-methylcytosine but cleaves with N^6-methyladenine.
[c] No cleavage with 4-methylcytosine or with 3′C of 5-methylcytosine.
[d] Cleavage with 3′C but not with 5′C of 5-methylcytosine.
[e] Active at 65 °C.

2.2 Design of the linker-adapter

The linker-adapter has been specifically designed to permit the amplification of DNA sequences to which it is ligated without itself becoming a substrate for amplification by PCR (*Table 1* and *Figure 2*). Several features are important in the design. Since the linker provides the primer binding site it is desirable that it be long enough to promote specific binding (a 20-mer primer seems ideal) but as short as possible to promote efficient ligation to all

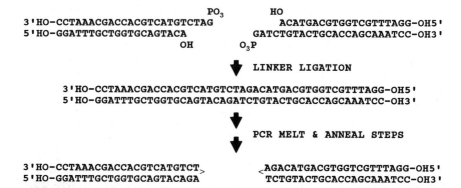

Figure 2. The major ligation product in the procedure outlined in *Figure 1* is the formation of linker dimers as shown above. Each strand of the linker dimer forms a hairpin duplex during the annealing step of PCR and thus is not a template in the polymerization step.

available target DNA compatible ends. The sequence of the 20-mer primer should be of approximately 50% GC content (providing annealing temperatures in the 52–60°C range during PCR) and should not contain homopolymeric sequences, particularly at its 3′-end, to avoid undesirable priming at such sequences which are common in the genome of complex eucaryotes. The primer should have no significant self-complementarity so as to discourage the formation of primer-dimer artefacts (primers priming DNA synthesis on primers) which may become significant competitor templates after many cycles of PCR. Only the 5′-end of the oligonucleotide component participating in the ligation reaction with the 3′-end of the target DNA is phosphorylated. This design feature limits the product of linker-with-linker ligation to the formation of dimers. Each strand of the resulting linker dimer is self-complementary, rapidly hybridizes with itself forming a hairpin duplex during the annealing step of PCR and thus is not a competing template in the amplification reaction (*Figure 2*). This is critical since the linker-with-linker ligation products are present at an enormous molar excess compared to the linker-modified target DNA. In most cases it is unnecessary and undesirable to include accessory restriction sites in the linker-adapter sequence for subsequent cloning of the amplified DNA. Restriction at these accessory sites has the potential to cause truncation of amplified sequences containing internal sites. Cloning of amplified DNA is discussed in Section 5.2.

3. Microdissection equipment

3.1 Microscope, micromanipulator, and vibration-free table

A standard Zeiss microscope (not an inverted model) equipped for phase contrast with 10×, 25×, and 40× plan-neofluar objectives is satisfactory for polytene chromosome dissection. An optivar may be added to increase magnification but with some loss of resolution. Oil immersion lenses may be employed for high-magnification, high-resolution work; however, these lenses have a very short working distance. Also, care must be taken to use a minimum of immersion oil to prevent admixing with the paraffin oil at the edge of the coverslip. This changes the refractive index of the immersion oil and destroys the phase contrast image. The microscope is ideally fitted with a circular rotating stage to permit proper positioning of the oil chamber with respect to the micromanipulator during microdissection (*Figure 3*).

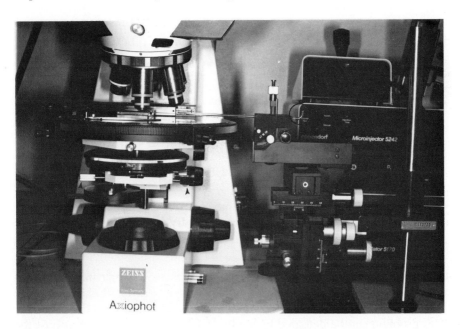

Figure 3. The microdissection theatre. Elements are described in the text.

After testing many models of manipulators, I prefer the motorized Zeiss instrument. It has very convenient coarse adjustments for initial positioning of micro-instruments in the oil chamber. Motorized motion with the joystick is precise, smooth and there is no drift as with some pneumatic instruments. A stand for the instrument can be fabricated by any machine shop or may be purchased as a custom item from Zeiss.

Most buildings vibrate too much to permit microdissection on a fixed table or bench. Therefore, the microscope and micromanipulator head holding the dissection instruments must be mounted on a vibration-free table (e.g. Micro-g). To prevent internal corrosion of the air pistons, the table is best operated using dry compressed air.

3.2 Oil chamber, glass instruments, and pipettor

The fabrication of the oil chamber and micro-instruments has been described in detail elsewhere (4, 5). The pipetting of nanolitre volumes in the oil chamber can be tedious to master since aqueous solutions resist extrusion from a micropipette into paraffin oil and must finally be made to cling from the underside of a glass coverslip during the procedure. The method outlined in *Protocol 1* for preparing instruments for micropipetting works well in my hands.

Protocol 1. Preparation of instruments for micropipetting

1. Prepare a fresh solution of 10% dimethyldichlorosilane (Fluka) in chloroform.

2. Rinse the glass instruments and coverslips (10 × 40 mm cut with a diamond pencil from 22 × 40 mm stock, Corning) in methanol and allow to dry in a dust-free place.

3. Dip the coverslips to hold the hanging drops and the microknives in the silane solution for several seconds and then immediately into methanol followed by distilled water.

4. Take micropipettes through this same series by aspirating the solutions in and out. Properly silanized glass emerges dry from the water rinse.

Placement of viscous aqueous solutions such as the enzyme and ligase mixes adjacent to the DNA-containing droplets can be difficult as these solutions may prefer to adhere to the pipette shaft rather than the coverslip. This problem can be defeated by gently scrubbing the pipette tip against the coverslip several times to roughen the surface at the point of placement of the drop followed by pipetting of the solution. An inexpensive pipettor which provides very fine control of solution flow into and out of the micropipettes is illustrated in *Figure 4*. The 10 cc plastic syringe is connected to a Y connector using silicone or latex tubing. One end of the Y connector is connected to the miropipette at the upturned glass nipple (circled, *Figure 3*) with a length of more-rigid tygon tubing. The other end of the Y connector carries a short length of thin-walled latex or silicone tubing that is open to permit the instant release of pressure on the micropipette or squeezed shut in order to apply pressure for pipetting the sample. Adroit manipulation of the open-ended latex tube permits very fine control of pipetting, particularly during phenol extraction where removal of the phenol phase must be performed without aspiration of the DNA-containing drop.

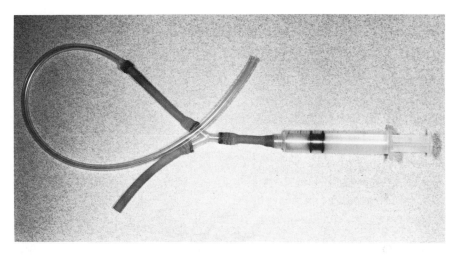

Figure 4. An apparatus for micropipetting. Features are described in the text.

4. Experimental procedures

4.1 Preparation of the linker-adapter

The unphosphorylated oligonucleotide components of the linker-adapter should be purified individually by electrophoresis on a 12% denaturing acrylamide gel to give single base resolution for purification (6). The duplex is then prepared as follows according to *Protocol 2*.

Protocol 2. Preparation of linker-adapter duplex

1. Elute the oligonucleotide from the acrylamide gel slice in a twofold excess of 10 mM Tris–HCl, pH 7.5, 1 mM Na$_2$ EDTA (TE) overnight at 4°C.

2. Concentrate the solution containing the oligonucleotide to approximately 0.5 ml by serial extractions with tertiary butyl alcohol.

3. Add 0.07 vol. of 3 M sodium acetate, pH 6.0, to the concentrated oligonucleotide and extract this solution with an equal vol. of phenol/chloroform (1:1) until the interface is clear (usually after one extraction).

4. Extract once with chloroform and precipitate with 2.5 vol. of ethanol at −20°C for several hours.

5. Recover the oligonucleotide by centrifugation at ~14 000 g in a microcentrifuge for 20 min, rinse with 70% ethanol, dry the pellet and redissolve in TE at approx. 1 mg/ml.

Protocol 2. *Continued*

6. Phosphorylate the oligonucleotide component of the linker-adapter carrying the 5′-end, involved in the ligation reaction with the target DNA, using T4 polynucleotide kinase as follows:
Mix:

- 10 µg oligonucleotide
- 7.5 µl 10 × kinase buffer[a]
- 3.5 µl 0.1 M dithiothreitol
- 3.5 µl 10 mM ATP
- 10 units T4 polynucleotide kinase
- Distilled water to a final vol. of 75 µl

Incubate the reaction at 37°C for 30 min.

7. Add 0.07 vol. of 3 M sodium acetate, pH 6.0, to the kinase reaction and extract once with an equal volume of phenol/chloroform (1:1) followed by chloroform.

8. Precipitate the phosphorylated oligonucleotide as in steps 4 and 5 above and redissolve in 20 µl of TE containing 10 µg of the complementary unphosphorylated (primer) component of the linker-adapter. Incubate this mixture at 58°C for 1 h to permit annealing of the linker duplex.

[a] 10 × kinase buffer: 750 mM Tris–HCl, pH 7.5, 1 M KCl, 0.1 M MgCl$_2$, 0.5 mg/ml acetylated BSA.

An alternative approach to linker-adapter synthesis which is slightly more expensive but has the advantage of assuring that all linker-adapters are phosphorylated appropriately is as follows:

Synthesize the palindrome resulting from linker-with-linker ligation (e.g. the 44-mer in *Figure 2*) resuspend in TE and anneal at 65°C for 2 h. Purify the 44-mer duplex by electrophoresis in a 12% non-denaturing acrylamide gel and recover the DNA from the gel slice as described above (*Protocol 2*, steps 1 to 5). Digest the duplex with the appropriate restriction enzyme to yield the linker-adapter.

4.2 Preparation of specimens, set-up of the oil chamber, and microscopic steps

The major consideration in preparing specimens is to avoid degradation or chemical modification of the DNA (4, 5). It is probably true that the preparation of metaphase chromosomes for microdissection remains to be optimized although workable methods have been developed (5). The method of Pirrotta and co-workers (4) has been used for work with polytene chromosomes as described in *Protocol 3*.

Protocol 3. Specimen preparation from Polytene chromosomes.

A. *Mounting and dissection.*

1. Use a diamond pencil to cut down 22 × 40 mm No. 1 coverslips (Corning) to 15 × 40 mm size. Dip the coverslips in methanol and wipe dry with a kimwipe tissue. Silanize 22 × 22 mm coverslips as described *Protocol 1*.

2. Place a 15 × 40 mm coverslip on a standard microscope slide with a glass chip removed at the top right corner and allow 15 μl of distilled water to seep by capillarity between the coverslip and the slide. Rapidly dissect salivary glands free of fatbody and squash them in 15 μl of 45% acetic acid with a silanized coverslip. Limit the exposure to acetic acid to one minute or less. Dip the slide/coverslip sandwich into liquid nitrogen until it stops boiling, then flip the silanized slip off with a razor blade at the edge. Submerge the slide with the chromosome-carrying coverslip frozen in place in 70% ethanol in a Copeland jar.

3. Transfer the coverslip, free of the slide, to 95% ethanol and soak for 10 min and then dry the specimen completely in a 37°C incubator. Place the coverslip on the oil chamber with the chipped edge now on the left (chromosome side down). Pipette paraffin oil (Baker, Photrex grade) between the coverslip and the bottom of the chamber until the space is filled. Place a 10 × 40 mm silanized coverslip next to but not touching the first coverslip on the oil chamber. Pipette paraffin oil saturated with distilled water[a] beneath the silanized cover slip. Use a permanent marker to draw a grid on the silanized coverslip.

4. Mount the oil chamber on the microscope stage with the silanized coverslip adjacent to the micromanipulator head as shown (*Figure 3*). Pressurize the vibration free table.

5. Using the 10 × objective, scan the chromosome coverslip for a suitable specimen. Focus the specimen at × 400. Use the knobs for centring the light source for Köhler illumination (arrows, *Figure 3*) to move the bright spot of illumination to the side of the field. Further adjustment of the phase rings or condenser should not be necessary if the microscope was previously tuned on microscope slide specimens. Move to the gridwork on the silanized coverslip and mount a micropipette in the oil chamber containing 10 mM Tris–HCl, pH 8.0, 10 mM NaCl, 0.1% SDS and 500 μg/ml proteinase K (Boehringer Mannheim). Pipette a one nanolitre drop (approximately 0.1–0.05 of the field) under each square of the grid and note the position of each on the stage vernier in a diagram of the grid.

6. Mount a microknife in the oil chamber and plunge the tip in and out of a DNA extraction drop to be sure that the drop does not adhere to the knife. Scrape the desired material from the specimen and deposit the material in the extraction drop noting the contents of each drop on the

Protocol 3. *Continued*

grid diagram. Incubate the oil chamber in a Petri dish with paraffin oil at 37°C for 1 h after addition of the last sample.

B. *Restriction and linker-adapters ligation*

7. Return the oil chamber to the microscope stage and place a drop containing a few hundred nanolitres of phenol (BRL, freshly extracted three times with 1 × restriction enzyme buffer) containing an equal volume of 1 × restriction enzyme buffer at a convenient site on the silanized coverslip using a micropipette. Phenol extract each DNA sample three times with approximately 4 vol. of phenol. Upon addition of phenol the DNA drop will detach from the coverslip and should be allowed to swim in the phenol for 3–5 min with each extraction cycle. Use a micropipette to supply phenol and to remove discard from the oil chamber.

8. Fill a micropipette with chloroform and carefully pipette a gentle cloud of chloroform at each DNA drop. Allow the chloroform to settle into the paraffin oil for 30 min. Place a 0.2 × size drop (relative to the DNA drop) of restriction enzyme (Boehringer Mannheim or BRL) and buffer (1:1 mixture of enzyme stock: 10 × buffer) next to each DNA drop and use a clean microknife to merge these droplets. Return the oil chamber to the Petri dish with paraffin oil and incubate at 37°C for 1 h or 65°C for 20 min for *Taq*I.

9. Allow the oil chamber to cool to r.t. in the Petri dish then return it to the microscope stage. Place an equal size drop of linker-adapter (50 µg/ml) and 0.25 × size drop of ligase mix[b] next to each DNA drop and use a clean microknife to merge these drops. Return the oil chamber to the Petri dish with paraffin oil and incubate at 6–10°C for 18 h.

10. Slowly add paraffin oil to the Petri dish until the oil chamber and coverslip are completely submerged by approximately 1 cm. Carefully slide the coverslip horizontally off the chamber, invert it and place it back on the chamber being careful that the coverslip remains submerged in paraffin oil during this manoeuvre. Under a stereo microscope, add 1 µl of TE to the ligation drops and permit them to mix for 30 min. Recover each drop into a 0.5 ml plastic test-tube for amplification of the linker-modified DNA by PCR.

[a] Paraffin oil saturated with distilled water: emulsify 4 parts paraffin oil with 1 part H_2O, place at 80°C overnight, then allow the phases to separate at room temperature for 24 h.
[b] Ligase mix: 0.25 M Tris–HCl pH 7.6, 50 mM $MgCl_2$, 5 mM ATP, 5 mM dithiothreitol, 25% polyethylene glycol 8000 mixed 1:1 with T4 DNA ligase (9.5 units/µl, Boehringer Mannheim).

4.3 PCR amplification of linker-modified DNA

Generally it is necessary to perform two sequential rounds of PCR (*Protocol 4*) with 20 to 25 cycles per round in order to recover a useful mass of DNA from a single microdissected polytene chromosome fragment encompassing approximately 200 kbp of the genome from *Drosophila melanogaster* larval salivary gland chromosomes.

Protocol 4. PCR amplification

1. To each linker-modified DNA aliquot add:

- 10 µl 10 × PCR buffer[a]
- 2 µl dNTP cocktail[b]
- 1 µl 20-mer primer component of the linker-adapter (0.2 µg/µl)
- 86 µl distilled H$_2$O
- 1 µl (5 units) *Taq* DNA polymerase (AmpliTaq, Perkin-Elmer/Cetus)

Mix and overlay with 75 µl paraffin oil.

2. Perform 25 cycles of PCR with 1 min 94°C denaturation, 2 min 52°C annealing and 3 min 72°C extension per cycle either in water baths or using a PCR cylcer (e.g. Perkin-Elmer/Cetus) with no delay between temperature shifts. The final extension cycle at 72°C is for 10 min.

3. Following the first round of PCR, dilute a 10 µl aliquot of the reaction into 90 µl of a fresh, appropriately concentrated reaction mix for an additional 15–25 cycles of PCR with a final extension cycle of 10 min.

4. Phenol extract and ethanol precipitate the product. Electrophorese a tenth of the product in a 1.2% agarose gel with *Hin*dIII cut lambda DNA as a size marker. The product generally appears as a smear of DNA from less than 100 bp to over 1 kbp.

[a] 10 × PCR buffer: 0.1 M Tris–HCl, pH 8.3, 0.5 M KCl, 15 mM MgCl$_2$, 0.01% (w/v) gelatin
[b] 10 mM dATP, dGTP, dCTP, dTTP; 0.1 M ultrapure stocks, Pharmacia

---

5. Labelling, cloning, and analysis of the PCR product

5.1 Labelling of the PCR product

It has been possible to isolate cosmid and YAC (yeast artificial chromosome) genomic DNA clones from *Drosophila melanogaster* using the microdissected-amplified-DNA itself as a labelled probe without prior fractionation (2, 7). The labelled PCR product may also be used to 'paint' the chromosome or dissected chromosome subregion by *in situ* hybridization (2; *Figure 5*).

The PCR product is first purified by ammonium acetate/ethanol precipitation or by gel fractionation to remove unlabelled nucleotides from the PCR reaction (6). An aliquot containing approximately 100 picograms of the PCR

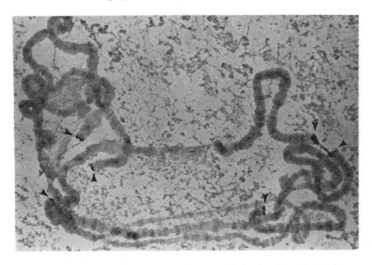

Figure 5. *In situ* hybridization of DNA microdissected from the tip of the X chromosome of *D. melanogaster* to larval salivary gland chromosomes. Arrowheads mark the points of hybridization on the telomeres of all chromosomes except those embedded in the chromocentre. DNA was labelled by PCR with biotin-II-dUTP and detected by strepavidin-HRP using diaminobenzidine as substrate.

product is then amplified in a $20 \mu l$ PCR reaction as described in *Protocol 4* but where biotin-11-dUTP ($50 \mu M$) or $\alpha\text{-}[^{32}P]dCTP$ ($10 \mu l$ of 3000 Ci/mmol, 0.1 mCi/ml, New England Nuclear) replaces dTTP or dCTP respectively. Biotin-labelled products are purified for *in situ* hybridization by ethanol precipitation. Radioactive products are purified by exclusion chromatography on a small column of p60 (Bio-Rad).

5.2 Cloning of the PCR product

It may be desirable to clone and analyse specific molecules from the amplified product. For material produced by *Mbo*I and *Sau*3A digestion of the target this is most easily accomplished by digesting the PCR product with the same enzyme and cloning the compatible sticky ended fragments into the *Bam*HI site of a plasmid vector (e.g. BlueScribe, Stratagene). Material produced by *Msp*I and *Hpa*II digestion of the target can be similarly cloned into the *Acc*I site of the same plasmid vector using *Msp*I or *Hpa*II to remove the linker-adapter. Note that it is necessary to remove the linker-adapters from the restricted product by gel fractionation (6) prior to cloning.

It is probably less desirable to clone blunt-ended fragments as this may lead to the cloning of many artifact products, especially from PCR reactions requiring many cycles of PCR. If blunt-end cloning is desired it is best to first blunt the ends of the PCR product using T4 DNA polymerase with only the 3'-most dNTP of the linker-adapter end (dCTP for linker-adapters in *Table 1*) in the reaction (6). Failure to copy the last base of a template or the addition of a non-template base with each amplification cycle as shown (8) has been found to be a frequent artefact of *Taq* DNA polymerase. The omission of the 3'- base is, however, only an arithmetic error since the 5'-end of each newly synthesized molecule is established by the primer.

5.3 Analysis of cloned PCR products

The analysis of cloned PCR products is most easily performed by PCR amplification of the cloned insert using primers, complementary to the plasmid sequences flanking the polylinker, which are normally used for sequencing of cloned inserts. Amplification of cloned inserts is performed according to *Protocol 5* for products cloned in BlueScribe.

Protocol 5. Amplification of BlueScribe inserts

1. Streak recombinants (white colonies on ampicillin/X-gal/IPTG plates) on to a master plate with a numbered grid. Allow the streaks to grow up.

2. Transfer a pinhead size aliquot of each colony with a toothpick to 10 µl of lysis buffer (10 mM Tris–HCl, pH 8.0, 50 mM Na_2 EDTA, 8% sucrose, 0.5% Triton X-100, 100 µg/ml lysozyme) in a 0.5 ml plastic test tube, vortex to mix, then place on ice (many samples may be accumulated at this step and processed simultaneously).

3. Transfer the tubes to a heated block or water bath at 98°C for 1 min then back on ice. Add 50 µl of TE to each sample, vortex, and spin in a microcentrifuge for 5 min. Store these samples frozen at −20°C.

4. Transfer a 0.5 µl aliquot of each supernatant to a 0.5 ml tube containing 9.5 µl of PCR cocktail (10 mM Tris–HCl, pH 8.3, 50 mM KCl, 1.5 mM $MgCl_2$, 0.01% gelatin, 200 µM of each dNTP, 2 µg/ml of each primer (eg. T3 and T7 17-mer sequencing primers for BlueScribe) and 50 units/ml *Taq* DNA polymerase (AmpliTaq, Perkin-Elmer/Cetus). Overlay each reaction with 20 µl of paraffin oil and amplify for 25 cycles as described in *Protocol 4*.

5. Electrophorese 5 µl of each reaction in 1.2–1.5% agarose gels for size determination and reverse-Southern blot analysis (9) if desired (*Figure 6*). Transfer the DNA to nylon membrane (e.g. GeneScreen Plus, DuPont) for hybridization experiments since these membranes efficiently retain small DNA molecules.

Individual cloned inserts may be labelled for use as hybridization probes in library screening or for *in situ* hybridization by the substitution of labelled nucleotides as in Section 5.1 above. The relatively small amount of plasmid

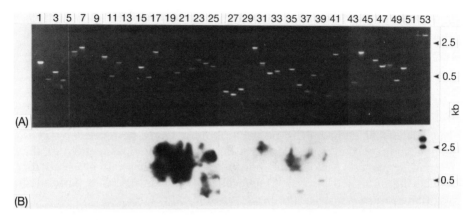

Figure 6. Fluorescent staining and reverse Southern blotting of microdissected DNA inserts from the 83D-F region of chromosome of *D. melanogaster*. PCR was used to amplify cloned inserts in plasmids isolated from a shot-gun-cloned library of amplified chromosomal DNA. The inserts were fractionated in a 1.5% agarose gel (A) and transferred to a nylon membrane which was hybridized with labelled total genomic DNA (B). **Lanes 1–50** potentially contained amplified, plasmid insert DNAs. **Lane 52** contained a 2.5 kb DNA fragment of *Drosphila* which is unique whole **lane 53** contained a similar-sized fragment which is repeated approximately 50 times per haploid genome. Clones from **lanes 18–25** were rescreened and appear to contain single-copy sequence. A longer exposure of the blot shows weak signals over all of the inserts whose intensity varies with the intensity of fluorescence, unlike clones in **lanes 29, 36,** and **39** which appear to contain genomically reiterated sequences.

sequences labelled in this procedure do not interfere with the unambiguous detection of the desired genomic sequences cloned in cosmid vectors (2).

6. Whole chromosome and whole genome PCR

For a variety of applications it may be desirable to produce a collection of amplifiable sequences from a whole genome or, for example, a specific human chromosome. Such a collection can then be labelled and use to 'paint' the chromosomes or segments of chromosomes by *in situ* hybridization for karyotype analysis (Tkachuk, Pinkel, Johnson, and Gray, in preparation). A collection of molecules could be assembled from microdissected chromosomes as described above or from DNA that has been purified from cells of an organism as the case may be. *Protocol 6* is a generally useful method that has been used for amplification of genomic DNA from *D. melanogaster*, DNA extracted from flow-sorted human chromosomes and DNA from cosmid and YAC clones.

Protocol 6. Whole genome or chromosome PCR

1. Digest 100 pg to 10 ng of the target DNA to completion with 0.1 unit of each of the enzymes shown in *Table 1* in 10 µl reactions for 1 h.

2. Place the digests on ice and add 4 µl of T4 ligase mix (*Protocol 3*, footnote *b*), 50 ng of the appropriate linker-adapter (*Table 1*), 5 units T4 DNA ligase (9.5 units/µl, Boehringer Mannheim) and water to 20 µl. Incubate the reaction at 4–6°C for 24 h.

3. Freeze the ligation mix at −20°C, thaw and add 1 µl of each ligation mix to 99 µl of ice-cold PCR reaction mix as for the amplification of microdissected material (*Protocol 4*). Amplify the material according to *Protocol 4*, step 2, for 20–30 cycles and test for amplification by gel electrophoresis of a 10 µl aliquot of the reaction in a 1.2% agarose gel so as not to exceed the capacity of the PCR reaction (approx. 1 µg).

References

1. Lüdecke, H.-J., Senger, G., Claussen, U. and Horsthemke, B. (1989). *Nature*, **338**, 348.
2. Johnson, D. H. (1990). *Genomics*, **6**, 243.
3. Saiki, R. K., Gelfand, D. H., Stoffel, S., Scharf, S. J., Higuchi, R., Horn, G. T., Mullis, K. B., and Ehrlich, H. A. (1988). *Science*, **239**, 487.
4. Pirrotta, V., Jäckle, H., and Edström, J.-E. (1983). *Genetic Engineering Principles and Methods*, **5**, 1.
5. Edström, J.-E., Kaiser, R., and Röhme, D. (1987). In *Methods in enzymology*, Vol. 151 (ed. M. M. Gottesman), p. 503. Academic Press, New York and London.
6. Maniatis, T., Fritsch, E. F., and Sambrook, J. (1982). *Molecular cloning: A laboratory manual*. Cold Spring Harbor Laboratory Press, Cold Spring Harbor, NY.
7. Garza, D., Ajioka, J. W., Carulli, J. P., Jones, R. W., Johnson, D. H., and Hartl, D. L. (1989). *Nature*, **340**, 577.
8. Johnson, D. and Henikoff, S. (1989). *Molecular and Cell Biology*, **9**, 2220.
9. Crampton, J. M., Davies, K. E., and Knapp, T. F. (1981). *Nucleic Acids Research*, **9**, 3821.

9

Inverse polymerase chain reaction

JONATHAN SILVER

1. Introduction

The PCR method requires the use of primers complementary to opposite strands of DNA and oriented so that primer extension proceeds 5' to 3' from each primer towards the other primer. The essential consequence of this configuration is that newly synthesized DNA strands contain a binding site for the opposite primer at their 3'-ends. Primer extension products can then serve as templates in the next round of DNA synthesis, and repeated cycles of melting, annealing, and primer extension lead to exponential amplification of the segment of DNA encompassing the two primers. The requirement for primers complementary to both ends of the segment to be amplified poses a problem when sequence is known only in one region, and one wishes to amplify DNA to one or both sides of that region. Three laboratories (1, 2, 3) independently recognized a simple solution to this problem: circularize a restriction fragment containing the region of interest, and then primers oriented so that primer extension proceeds 'outward' from the region of known sequence are correctly positioned to amplify the flanking DNA (see *Figure 1*). This trick of circular permutation has been dubbed 'inverse' or 'inside-out' PCR. It has been used to amplify DNA flanking integrated retroviruses (1), bacterial insertion elements (2), a segment of a malaria gene (3), and the ends of inserts in cloning vectors such as yeast artificial chromosomes (4). Other possible uses include extending partial cDNA clones in the 3' or 5' direction, amplifying the promoter region upstream of the start site of a mRNA, and chromosomal walks in genomic DNA.

2. Theoretical considerations

2.1 Circularization

The simplest approach to inverse PCR is to cut double-stranded target DNA with an appropriate restriction enzyme, dilute the DNA and religate to form circles (see *Protocol 1*). Several considerations will affect the choice of restriction enzyme. First, it should generate a fragment of appropriate size for circularization and amplification. Fragments shorter than 200–300 bp are

137

Inverse polymerase chain reaction

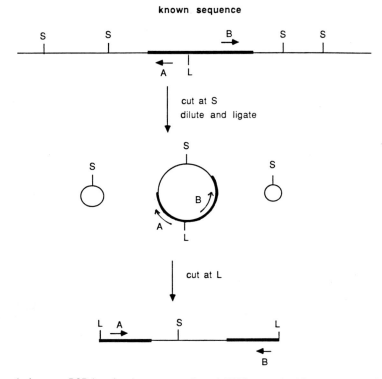

Figure 1. Inverse PCR by circular permutation. A DNA strand with a segment of known sequence (heavy line) is cut with a restriction enzyme S, diluted and ligated to form circles. Circles containing the region of known sequence plus flanking DNA are linearized by cutting with a restriction enzyme which cuts at site L. Primers A and B (5'→3' orientation indicated by arrows) are then properly oriented to amplify flanking DNA.

difficult to circularize (5), and fragments larger than about 3 kb are, at present, difficult to amplify from complex mixtures. Restriction enzymes which leave four base overhangs may facilitate subsequent ligation. The choice of restriction enzyme will obviously be affected by information about restriction sites in the region of known sequence and flanking DNA. For example, if the region of known sequence is small and it is desired to amplify flanking DNA on both sides, the enzyme must not cut within the region of known sequence. If DNA to only one side of the known sequence is to be amplified, the enzyme must cut within the region of known sequence but far enough from its boundary to leave binding sites for two primers in the segment to be amplified. If a restriction map of the region of known sequence and its flank is not available or easily obtainable, it may be appropriate to choose a restriction enzyme with a four base recognition sequence. Partial and/or complete digests with such an enzyme are likely to generate fragments which are not too large to amplify and which contain a useful portion of the known sequence plus flanking DNA. Another consideration is the availability of a restriction site to linearize circularized DNA at a position which will not interfere with subsequent PCR, i.e. between the 5' ends of the primers (site L, *Figure 1*). Linearization increases sensitivity and may be important if the amount of template is limiting (see Section 2.2 below).

After digestion, the restriction enzyme is inactivated by heat or phenol extraction, and circularization is performed by ligating at relatively low DNA concentration. Low concentration of DNA favours circularization over con-catamerization. A random coil model of DNA predicts that for 1 kb molecules at a concentration of 1 μg/ml, the local concentration of one end of a molecule in the vicinity of its other end, is approximately 60 times higher than the concentration of ends of other molecules (6); thus, intra-molecular ligation is more likely than inter-molecular ligation. The relative likelihood of ligation of one end of a molecule to the other end of the same molecule, as opposed to another molecule, is given by the expression $1900/c(bp)^{1/2}$, where c is concentration in μg/ml and bp is length in base-pairs (6). For values of c and bp for which this ratio is $\gg 1$, almost all of the molecules should, in theory, circularize. Little would be gained by further dilution, and in fact it might be advantageous to ligate at somewhat higher concentration so that the final concentration of circular template would be greater, and a concentration step could be avoided (see 'short-cuts', Section 3.2). If the DNA to be amplified is complex (e.g. genomic DNA), then concatamers do not seriously interfere with inverse PCR, since they amplify arithmetically rather than exponentially. On the other hand, if the DNA is of low complexity (e.g. plasmid DNA), then concatamers can lead to spurious amplification products when two fragments containing primer binding sites are joined together in the appropriate orientation.

Variations on the circularization theme are possible. For example, to extend cDNA clones in either direction, double-stranded cDNA can be synthesized and ligated to linkers containing rare restriction sites. Cleavage and ligation at low DNA concentration then leads to appropriate templates for inverse PCR. The same approach could, in theory, be used to 'walk' outward from a segment of genomic DNA: synthesize double-stranded cDNA starting with a specific primer, using random hexamers to form the second strand, then ligate linkers, cleave and religate to form circles. This approach would, in theory, permit the 'walk' to go an arbitrary distance into the flank, rather than relying on the chance location of neighbouring restriction sites.

2.2 Linearization

The need for this step is not universally accepted. In preliminary experiments, we found that linearization of covalently closed circular double-stranded DNA increased the efficiency of PCR more than 100-fold, as determined by serial dilution of template. A likely explanation is that covalently closed circular double-stranded DNA is a poor template for the first round of primer extension because of 'snap-back' reannealing of the entwined DNA strands. In contrast to double-stranded circles, single-stranded circles and nicked double-stranded circles are excellent templates for PCR. Hence, if the template for inverse PCR is present in limiting amounts and is almost all in the form of covalently closed double-stranded circles, linearization or nicking may be necessary. If linearization is done by restriction enzyme digestion, the enzyme must not cut between the 3'-ends of the primers on the circularized template, or it would destroy the template for PCR. This can pose a problem when a restriction map of the flanking region is not readily available. In this case one may take a statistically favourable approach, namely, use an enzyme which has a six-base recognition site to linearize, in combination with an enzyme which has a four-base recognition sequence to make the original fragments for circularization. Only about 1 in 16 such circularized fragments should contain a site for the six-base cutting enzyme between the 3'-ends of the oligos, i.e. where it could interfere with amplification. An alternative to cutting with a restriction enzyme is to randomly nick the circularized fragments, for example by heating (7) or very mild DNAse treatment; this approach has the potential drawback of destroying many templates if the nicking is too extensive. Another approach, which is the subject of current research, is to circularize single-stranded primer extension products to serve as templates for inverse PCR.

2.3 Amplification

Once the circularly permuted template is made, 'inverse' PCR is no different from 'normal' PCR, and the same considerations apply to choice of primers and PCR conditions. These issues are discussed in detail in other chapters of this book.

3. Practical aspects of inverse PCR

3.1 Basic procedure

Protocol 1. Circular permutation of template DNA

1. Digest target DNA with an appropriate restriction enzyme; confirm adequate digestion by gel electrophoresis.
2. Phenol extract, ethanol precipitate, resuspend in TE buffer (10 mM Tris–HCl pH 8.0, 1 mM EDTA).
3. Dilute DNA to 1–10 μg/ml in ligase buffer,[a] add 1 unit T4 DNA ligase and incubate at 14°C overnight.
4. After ligation, phenol extract, ethanol precipitate, and resuspend in TE buffer.

5. Dilute in buffer appropriate for the restriction enzyme to be used for linearization and digest with this restriction enzyme. Confirm digestion of controls by gel electrophoresis, phenol extract, ethanol precipitate and resuspend in TE buffer.

6. Proceed to PCR.

^aLigase buffer: 50 mM Tris–HCl pH 7.6, 10 mM MgCl₂, 1 mM ATP, 1 mM DTT

Controls may include a fragment of cloned DNA comparable in size to the intended target and prepared and diluted in the same way as the DNA containing the target. Circularization of the control may be assessed by gel electrophoresis in the presence and absence of ethidium bromide (see below).

3.2 Short-cuts

It may be possible to simplify the *Protocol 1* significantly by eliminating most or all of the phenol–ethanol steps, provided the enzymes used can be heat inactivated and will work in T4 ligase buffer or *Taq* polymerase buffer and supplemented with ATP. This can be checked in pilot experiments using plasmid DNA. For example, we found that *Sau*3A works in *Taq* buffer (10 mM Tris–HCl pH 8.3, 50 mM KCl, 2.5 mM MgCl$_2$, 0.01% gelatin) and that *Pst*I and T4 DNA ligase work in *Taq* buffer supplemented with ATP to 67 µM. This amount of ATP did not interfere with subsequent PCR. Thus, in appropriate circumstances, *Protocol 1* can be simplified to *Protocol 2*.

Protocol 2. Quick circular permutation and inverse PCR

1. Digest target DNA with an appropriate restriction enzyme in *Taq* buffer (1 h) and heat inactivate restriction enzyme at 68°C, 10 min.

2. Dilute DNA to 1 µg/ml, add ATP to 67 µM and 1 unit T4 DNA ligase. Ligate at 14°C, overnight, heat inactivate the ligase at 68°C, 10 min.

3. Linearize with an appropriate restriction enzyme in the same buffer (1 h).

4. Add amplifying primers (50 pmol each), dNTPs to 0.2 mM, 2.5 units *Taq* polymerase and commence thermo-cycling.

---

4. Amplification of cellular DNA flanking integrated retroviruses

4.1 Theoretical considerations

Retroviruses have a repeated sequence of several hundred nucleotides at either end of the integrated, or 'proviral', DNA form. If primer sequences are chosen from this long terminal repeat (LTR), two segments of DNA will be amplified, one from the 'left-hand' or '5''' LTR (the site of initiation of viral transcripts), and one from the 'right-hand' or '3''' LTR. The position of the inverse PCR primers with respect to the cut site in the LTR determines whether the 5' or 3' flank will be amplified (see *Figure 2*). In either case, one of the LTRs gives rise to an amplified fragment containing flanking cellular DNA, and the other LTR gives rise to a fragment consisting entirely of retroviral sequence; the latter can serve as an internal control. Because many copies of retrovirus-like elements are inherited, it is important to choose primers which are specific for the targeted retrovirus.

4.2 Results

Figure 2 shows the strategy we used to amplify the 5' flank of an inherited murine retrovirus, and *Figure 3* shows the results. In addition to amplifying the correct fragments 'upstream' of the 5' and 3' LTRs, the PCR reaction generated large amounts of 'primer-dimers' and other artefactual fragments; this is probably a result of the large number of thermocycles performed in this

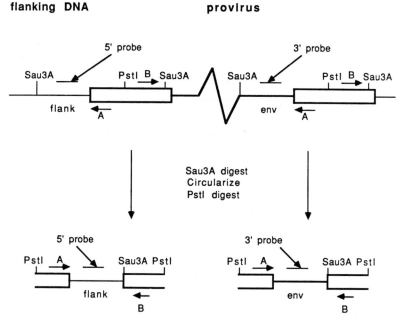

Figure 2. Inverse PCR strategy applied to amplifying DNA 5' of the LTRs of an integrated retrovirus. Flanking cellular DNA (flank) indicated by thin lines, provirus by thick line, LTRs by rectangles. env = retroviral envelope gene. Primer sequences: A, 5' ATGCGGCCGCGTCGACCTGGCTAAGCCTTATGAA 3'; B, 5' ATGCGGCCGCGAATT-CCCCAGATGACCGGGGATC 3'. These primers have artifical *Not*I and *Sal*I (A) or *Not*I and *Eco*RI (B) sequences at their 5'-ends to facilitate subsequent cloning.

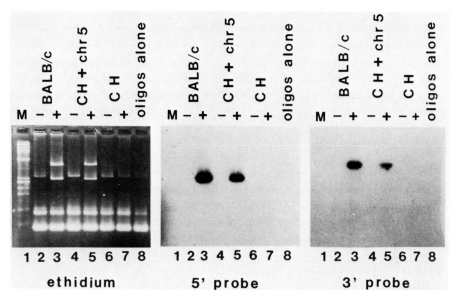

Figure 3. Inverse PCR of cellular DNA flanking the endogenous ecotropic virus of BALB/c mice. **Left panel**: ethidium-bromide-stained agarose gel. M, φX *Hae*III molecular-weight markers; − = DNA was cut by *Sau*3A but not circularized; + = DNA was cut by *Sau*3A, diluted to 1 μg/ml, ligated, and cut with *Pst*I. BALB/c, BALB/c mouse DNA; CH + chr 5, DNA from Chinese Hamster-BALB/c mouse somatic cell hybrid containing mouse chromosome 5 as its only mouse chromosome; the BALB/c endogenous ecotropic virus has been mapped to mouse chromosome 5; CH = Chinese Hampster DNA; oligos alone, no template DNA was added to PCR. Thermocycling conditions were 94°C × 45 sec, 50°C × 45 sec, 72°C × 4 min for 60 cycles. **Middle panel**: Southern blot of gel shown at left was hybridized to radiolabelled oligonculeotide probe for cellular DNA flanking BALB/c ecotropic virus (sequence 5′TGTCTTCATTGGGTACGCGG 3′). **Right panel**: same blot was stripped and rehybridized to oligonucleotide probe for 3′ region of BALB/c ecotropic virus env gene (sequence 5′ AATAAAAGATTTTATT 3′). (This figure is reproduced from ref. 1 with permission.)

experiment (see legend to *Figure 3*) and the fact that one of the primers terminated with the palindromic sequence GATC (see 'Artefacts', Section 6 below). With hindsight, it would have been better to have terminated the PCR reaction after 35–40 cycles, and to have used a primer without a terminal palindromic sequence.

5. Trouble-shooting

It is useful, when starting, to have a model system to test whether various steps in the procedure are working. The model system may consist of a cloned DNA containing the primer binding sites, as well as relevant restriction enzyme sites to allow circular permutation. The fate of such a molecule can be monitored more simply than that of the real target for PCR, and the surrogate DNA can be diluted to single copy amounts in carrier DNA to test sensitivity. When working with such a model system, it is very important to avoid contamination of reagents with cloned or PCR amplified DNA. Since this may happen despite all precautions, it is extremely important to design the model template so that amplification products derived from it, can be easily distinguished (e.g. by size) from the legitimate product one wishes to amplify.

Circularization is probably the most crucial and difficult step in inverse PCR, and determining whether it has occurred is not easy in complex mixtures. One approach to monitoring circularization is to dope the starting DNA with a surrogate radiolabelled restriction fragment comparable in size to that of the target. The mobility of this fragment in agarose gel electrophoresis can be monitored before and after ligation, and in the presence and absence of ethidium bromide. Ethidium bromide increases the mobility of covalently closed circles and decreases the mobility of linears. However, this test is not ideal as it provides only circumstantial evidence for circularization of the actual target DNA.

Careful consideration should be given to how PCR products will be tested to see if they are correct. Artefacts do occur (see Section 6) and often much more time is spent in trying to determine what has been amplified than in designing and carrying out the amplification in the first place. Negative controls often show that an amplification product is incorrect. Such controls should include PCR reactions with no template, unrelated template, uncut template, uncircularized template, and covalently closed (unlinearized) circular template. Positive controls with model templates are worthwhile but should be done in separate experiments as they are a potential source of contamination. For a positive control to make sure the *Taq* polymerase, buffers, and thermocycler are functioning properly, it is safer to amplify an unrelated fragment.

In choosing primers and restriction enzymes for inverse PCR, it may be helpful to synthesize additional, 'nested' primers. Note that in inverse PCR, 'nested' primers are located 'outside' of the primers used for initial amplification, when viewed from the perspective of the original, uncircularized DNA. The 'nested' primers can be used to reamplify the products of the first PCR to enhance the sensitivity and specificity of the reaction (see Chapter 4). Nested primers are also useful to reamplify fragments for sequencing and to make hybridization probes to check the specificity of amplified DNA.

6. Artefacts

'Primer-dimer' artefacts are frequent in all kinds of PCR reactions. They may be reduced by avoiding primers which self-anneal or anneal to other primers in the same PCR reaction, by limiting cycle numbers, minimizing cycle lengths, increasing 'annealing' temperatures, and decreasing concentrations of Mg^{2+}, *Taq* polymerase, and primers.

Primer dimers may have an additional adverse effect in inverse PCR, namely they may lead to amplification of the region between the 5'-ends of the primers. At first site this seems paradoxical since *Taq* polymerase extends primers in the 3' direction. However, the 3'-end of each strand of a primer-dimer contains the reverse complement of one of the primers, and if this end of the primer-dimer binds to its complement on the template DNA, its 3' extension product copies the sequence lying 5' of the primer to which it binds. Consistent with this possibility, one of the products of inverse PCR which we cloned and sequenced consisted of the portion of a retroviral LTR which lies between the 5'-ends of the primers, flanked on either end by a primer-dimer. In this artefact, the primer-dimer portion consisted of the sequence of one primer, then three Gs, then the reverse complement of the other primer, raising the possibility that the primer-dimer arose after non-template directed addition of G or C residues to the ends of the primers.

Another artefact we observed in amplifying flanks of integrated retroviruses in tumour DNA was a smear or series of faint bands which could be eliminated by digestion with a single strand specific nuclease (mung bean nuclease). These at least partially single-stranded products may have arisen from the annealing of amplified DNA strands which shared homology over only part of their length. When DNA fragments containing LTR plus flanking cellular DNA, are amplified from several integrated retroviruses in the same DNA sample, the LTR sequences will be common to all the fragments, but the flanking of cellular DNA portions will be different for each fragment. Melting and reannealing of such fragments would give rise to partially double-stranded, partially single-stranded hybrid molecules. Consistent with this hypothesis, treatment of such samples with mung bean nuclease led to the disappearance of faint bands or smear, and appearance of a new ethidium-staining fragment whose size was that of the portion of the LTR which would be common to all of the fragments. Similar artefacts may arise in standard PCR reactions whenever multiple fragments are amplified which share homology over only part of their length.

7. Comparison with other methods

Inverse PCR is related to hemi-specific PCR methods which rely on one specific primer, and one non-specific primer, the binding site for which is attached by tailing or ligation (7, 8, 9). These hemi-specific methods allow amplification of sequences which lie to one side of a region of known sequence. The hemi-specific methods have an advantage over inverse PCR in that they do not require circularization. On the other hand, hemi-specific methods are, in theory, less powerful than inverse PCR because of their reliance on one relatively non-specific primer binding site in each PCR reaction. Most hemi-specific methods have been used on DNAs of reduced complexity, such as cDNA (see Chapter 10), although one recent report describes the application of a related method to genomic DNA (10). Improvements in inverse PCR will probably come from improved methods of circularization and elimination of the need to nick or linearize circularized templates.

Acknowledgements

The author thanks Vijaya Keerikatte and Franz Sels who, as graduate students, helped to develop these techniques in my laboratory.

References

1. Silver, J. and Keerikatte, V. (1989). *Journal of Virology*, **63**, 1924.
2. Ochman, H., Gerber, A. S., and Hartl, D. L. (1988). *Genetics*, **120**, 621.
3. Triglia, T., Peterson, M. G., and Kemp, D. J. (1988). *Nucleic Acids Research*, **16**, 8186.
4. Ochman, H., Medhora, M. M., Garza, D., and Hartl, D. L. (1990). In *PCR protocols, A guide to methods and applications* (ed. M. A. Innis, D. H. Gelfand, J. J. Sninsky, and T. J. White), pp. 219–27. Academic Press, San Diego, California.
5. Shore, D., Langowski, J., and Baldwin, R. L. (1981). *Proceedings of the National Academy of Sciences, USA*, **78**, 4833.
6. Maniatis, T., Fritsch, E. F., and Sambrook, J. (ed.) (1982). *Molecular cloning: A laboratory manual*, p. 286. Cold Spring Harbor Laboratory, Cold Spring Harbor, NY.
7. Frohman, M. A., Dush, M. K., and Martin, G. R. (1988). *Proceedings of the National Academy of Sciences, USA*, **85**, 8998.
8. Loh, E. Y., Elliot, J. F., Cwirla, S., Lanier, L., and Davis, M. M. (1989). *Science*, **243**, 217.
9. Ohara, O., Dorit, R. L., and Gilbert, W. (1989). *Proceedings of the National Academy of Sciences, USA*, **86**, 5673.
10. Mueller, P. R. and Wald, B. (1989). *Science*, **246**, 780.

10

PCR-directed cDNA libraries

SARAH JANE GURR and MICHAEL J. McPHERSON

1. Introduction

Construction of a PCR-directed cDNA library from total RNA may provide the only methodological approach to analyse cell-specific gene expression where the amount of biological tissue is severely restricted. This approach is particularly applicable where a specific stimulus results in the modification or differentiation of a small number of cells within a population. We are interested in studying changes in gene expression in plant roots where some 0.1 to 0.01% of the tissue is modified by an invasive pathogen. The library used as an example in this chapter was constructed from 1 μg of RNA derived from dissected plant tissue of which approximately 1% was the differentially modified material of interest (1). This chapter describes protocols for the construction and differential screening of such PCR-directed cDNA libraries. Similar strategies have been used successfully for library construction from animal cells (2).

1.1 Overview of the strategy

Schemes for the construction of cDNA libraries require the isolation of cellular RNA and usually further purification of mRNA from the more abundant rRNA and tRNA components. The major advantage in PCR-based methods is that mRNA purification is not necessary; indeed often where biological material is limited such mRNA purification would be impossible. The general strategy is illustrated schematically in *Figure 1*.

An oligo-dT primer and the enzyme reverse transcriptase are used to produce first-strand cDNA corresponding to the polyadenylated mRNA population. Excess oligo-dT primer is removed by selective precipitation or ultrafiltration to prevent interference in subsequent steps. A homopolymer tail, composed of dG residues, is added to the 3′-end of the first-strand cDNA then the template RNA is hydrolysed by alkali treatment. Second-strand cDNA synthesis by *Taq* polymerase, during the first cycle of PCR, uses an oligo-dC primer complementary to the newly added homopolymer dG tail. Subsequent cycles use both oligo-dT and oligo-dC primers to amplify the cDNA products. These primers also include restriction sites at their 5′-ends to

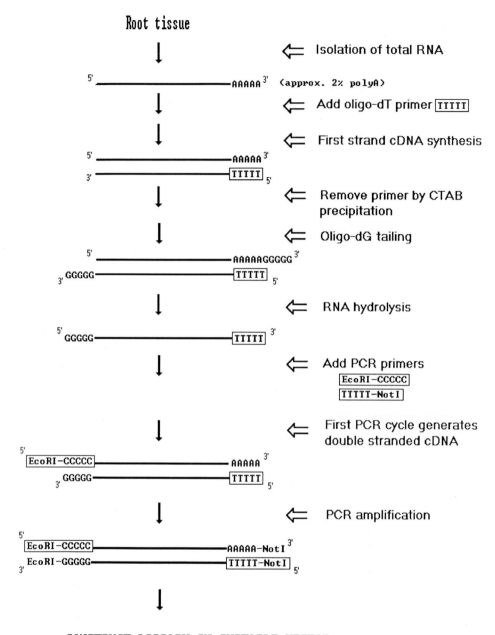

Figure 1. Schematic illustration of the PCR-based method for construction of representative cDNA libraries.

facilitate subsequent cloning; in addition these 5'-ends provide greater specificity during the annealing steps of the PCR than homopolymer primers alone.

Our strategy utilizes (a) different annealing temperatures during the early PCR cycles to ensure good priming and (b) size fractionation to prevent preferential cloning of very small products. There is an inherent bias favouring the amplification of small products which rapidly come to dominate the reaction. It is therefore essential to recover larger products to ensure good representation of long or full-length clones.

Protocols are provided for cDNA production and amplification together with those for library construction and differential screening. The approach is illustrated by an example library derived from plant material although the procedures are suitable for any eukaryotic system.

2. cDNA preparation

2.1 Isolation of total RNA

The first step towards construction of a large and representative cDNA library is the preparation of high quality, intact RNA. We have compared two methods that employ powerful denaturants, the first using guanidinium isothiocyanate (2) and the second guanidinium hydrochloride (3). The latter is our preferred method (see Section 2.2) and is described in *Protocol 1*. Harvest the tissue into foil packets and freeze immediately in liquid nitrogen. Crush the packet gently with a pestle to powder the tissue and store at $-70°C$. A more detailed discussion of the isolation of nucleic acids from plant material is provided in ref. 3. Alternative RNA preparation methods may be used for other sources of tissue.

Important notes. For all procedures in the preparation and manipulation of RNA it is essential to pre-treat glassware and disposable plastics in 0.02% diethylpyrocarbonate (DEPC) and bake for 72 h at 186°C or autoclave respectively. Wear gloves at all times to avoid nuclease contamination problems!

Protocol 1. Preparation of total RNA

1. Pre-cool disposable plastic 20 ml Sarstedt tubes in liquid nitrogen.

2. Transfer up to 100 mg pre-crushed tissue to the tubes and add 0.5 ml GuHCl buffer[a] and 0.5 ml phenol/CH_3Cl/isoamyl alcohol (25:24:1) (8).

3. Allow the tubes to thaw on ice and homogenize the samples by 15 strokes of a Polytron blender (Kinematica) returning the samples to ice after grinding.

4. Separate aqueous and phenol phases by centrifugation, 1000 g at r.t. for 10 min in a bench-top centrifuge.

Protocol 1. *Continued*

5. Remove the upper aqueous phase to microcentrifuge tubes avoiding any interface material. Re-extract with phenol, vortex briefly, and centrifuge at ~13 000 g at r.t. for 10 min in a microcentrifuge.

6. Repeat step 5 until the interface between the aqueous and phenol phases is clean (usually three extractions with plant tissue).

7. Add 0.2 vol. 1 M acetic acid and 0.7 vol. cold absolute ethanol to the aqueous phase and precipitate at −20°C overnight. The precipitate may not become visible immediately.

8. Pellet the RNA in a microcentrifuge at 4°C, 10 min. Wash the pellet by vortexing in 400 µl 3 M sodium acetate, pH 5.5, at 4°C to remove low molecular weight RNA species and polysaccharides.

9. Pellet the RNA and repeat step 8. Remove the salt with a final 70% cold ethanol wash.

10. Dissolve the pellet in 30–50 µl DEPC-treated H_2O. If the pellet is slow to dissolve, heat sample to 95°C for 1 min, vortex, and rapidly cool on ice.

11. Store the RNA at −70°C.

*"GuHCl buffer: 8 M guanidinium–HCl, 20 mM Mes (4-morpholino-ethanol-sulphonic acid), 20 mM Na₂EDTA, adjust to pH 7 with NaOH, add 50 mM 2-mercaptoethanol prior to use.

Assess the integrity and quantity of the prepared RNA by

(a) comparing $OD_{260/280}$ ratios (good RNA should give a value of about 2);

(b) electrophoresis through 0.66 M formaldehyde gels containing 0.5 µg/ml ethidium bromide (4): bands corresponding to ribosomal RNA sub-units (28S and 18S) should be clearly visible against a faint smear of mRNA.

Where material is limiting such analyses will probably not be possible. In such cases a more abundant tissue may be treated in parallel and used as an indicator of RNA quality.

2.2 First-strand DNA synthesis

The first-strand cDNA provides the template for second-strand synthesis and so it is important that its length and integrity are optimized. For this reason it is worth assessing the efficiency of first-strand cDNA synthesis by including $\alpha-[^{32}P]$ dCTP. We do this as described in *Protocol 2* by setting up two parallel reactions. One reaction is unlabelled and is used for library construction whilst a second reaction is labelled allowing the products to be visualized after alkaline gel electrophoresis (*Protocol 2B*).

Protocol 2. First-strand cDNA synthesis from total RNA

A. *Synthesis reaction*

Set up two identical reactions but to one add $10\,\mu\text{Ci}$ $[\alpha^{-32}\text{P}]$ dCTP (3,000 Ci/mmol). This is the *labelled* reaction used in part B of this Protocol.

1. Denature $1\,\mu\text{g}$ total RNA for 5 min in a $70\,^{\circ}\text{C}$ water bath and rapidly cool on ice.

2. Combine the following components, in order, to give a total volume of $20\,\mu\text{l}$:

   - H_2O to $20\,\mu\text{l}$
   - $5 \times$ AMV-RT buffer[a] $4\,\mu\text{l}$
   - Oligo (dT_{17}) $100\mu\text{g/ml}$ $2\,\mu\text{l}$
   - 10 mM dNTP $2\,\mu\text{l}$
   - 10 mM spermidine–HCl $1\,\mu\text{l}$
   - 80 mM sodium pyrophosphate $1\,\mu\text{l}$
   - Human placental ribonuclease inhibitor (HPRNI)   25 units
   - Total RNA $1\,\mu\text{g}$
   - AMV reverse transcriptase   10 units

3. Incubate at $42\,^{\circ}\text{C}$ for 1–2 h.

4. Add $20\,\mu\text{l}$ 0.1 M NaCl, 40 mM EDTA. Freeze the unlabelled reaction mix at $-20\,^{\circ}\text{C}$. Proceed to part B with the labelled mix.

B. *Assessing first-strand synthesis*

5. To the labelled reaction mix from step 4, add $20\,\mu\text{l}$ carrier DNA (salmon sperm, $100\,\mu\text{g/ml}$ in 20 mM EDTA) and $14\,\mu\text{l}$ 2 M NaOH.

6. Incubate at $96\,^{\circ}\text{C}$ for 30 min to hydrolyse the RNA.

7. Add $14\,\mu\text{l}$ 1 M HCl, $14\,\mu\text{l}$ 1 M Tris–HCl (pH 8.3) and an equal vol. ($\sim 80\,\mu\text{l}$) of Tris-saturated phenol/chloroform (1:1; v/v). Vortex then spin in a microcentrifuge, 1 min.

8. Remove the supernatant and extract with an equal vol. of chloroform, spin in a microcentrifuge, 1 min.

9. To the supernatant add an equal vol. of 4 M ammonium acetate. To this add 2 vol. cold ($-20\,^{\circ}\text{C}$) ethanol. Precipitate on dry ice for 15 min or preferably overnight at $-20\,^{\circ}\text{C}$.

10. Warm to room temperature (to dissolve precipitated dNTPs) and pellet in a microcentrifuge, 10 min.

11. Carefully remove the supernatant and wash the 'pellet' (which probably will not be visible) in $50\,\mu\text{l}$ 2 M ammonium acetate and $100\,\mu\text{l}$ cold ethanol. Spin in a microcentrifuge at r.t., 10 min.

Protocol 2. *Continued*

12. Wash the 'pellet' in 200 µl cold ethanol and spin in a microcentrifuge, 2 min.

13. Prepare a 1.4% agarose gel in 50 mM NaCl, 1 mM EDTA. Equilibrate the gel for 30 min in 1 × running buffer (30 mM NaOH, 10 mM EDTA).

14. Resuspend the labelled cDNA 'pellet' in a suitable volume of alkali loading buffer[b] and load on gel. Also load end-labelled molecular weight markers (4).[c] Electrophorese at a maximum of 7.5 V/cm until the Bromophenol Blue dye has migrated two-thirds of the gel length.

15. Soak the gel in two changes of 7% trichloroacetic acid for 30 min each. Transfer gel to Whatman 3MM paper and cover with Saran wrap (Dow Chemicals).

16. Dry on a slab gel dryer overnight with heating (60°C) and expose at −70°C to X-ray film (preferably pre-flashed) in a cassette with intensifier screens for approximately 12 h.

[a] 5 × AMV-RT buffer: 250 mM Tris–HCl, pH 8.3, 250 mM KCl, 50 mM MgCl$_2$, 5 mM DTT, 5 mM EDTA, 50 µg/ml bovine serum albumin (BSA)
[b] Alkaline loading buffer: 50 mM NaOH, 1 mM EDTA, 2.5% Ficoll 400, 0.025% Bromocresol Green, 0.025% Bromophenol Blue.
[c] DNA size markers are usually one of the following (a) λDNA *Hin*dIII digest, (b) ϕX174 *Hae*III digest, (c) a 123 bp ladder (BRL).

We obtain cDNA molecules ranging in size from approximately 200 bp to 2 kb with smaller amounts of significantly larger products. Of the two RNA preparation methods compared (see Section 2.1) the GuHCl method (3) resulted in the synthesis of longer products.

2.2.1 Alternative cDNA synthesis procedure

We have also prepared first-strand cDNA with Moloney murine leukaemia virus (MLV) reverse transcriptase from Pharmacia under the buffer conditions recommended by the manufacturer. A protocol for the use of the MLV enzyme in cDNA synthesis is provided in Chapter 13 of this book.

2.3 Removal of oligo-dT primers

The need to remove the oligo-dT primers following first-strand cDNA synthesis cannot be overemphasized. The addition of dG residues to residual oligo-dT primers during the homopolymer tailing reaction would provide spurious priming sites to which the amplimers would anneal during the PCR reactions. This would result in highly efficient amplification of small artefactual primer-based products at the expense of cDNA amplification. *Protocol 3* describes an efficient selective precipitation procedure involving cetyltrimethylammonium bromide (CTAB; 2) which precipitates the RNA:cDNA hybrids whilst leaving the primers in solution.

Protocol 3. Removal of primers

1. Thaw the unlabelled reaction mix from *Protocol 2A*. Add $0.5\,\mu g$ poly I· poly C (Pharmacia) and $3\,\mu l$ 10% (w/v) CTAB. Spin in a microcentrifuge at r.t., 20 min.

2. Carefully pipette-off supernatant and resuspend the 'pellet' (which may not be visible) in $14\,\mu l$ 1 M NaCl.

3. Add $25\,\mu l$ H_2O and $1\,\mu l$ 10% CTAB and spin in a microcentrifuge at r.t., 20 min.

4. Resuspend pellet in $10\,\mu l$ 1 M NaCl and reprecipitate with 2.7 vol. ethanol overnight at $-20\,°C$.

5. Spin in a microcentrifuge at $4\,°C$, 20 min. Carefully remove supernatant and wash 'pellet' in 70% ethanol. Spin in a microcentrifuge at $4\,°C$, 20 min, and briefly vacuum dry 'pellet'.

2.4 Oligo-dG tailing

Addition of a number of dG residues to the 3′-end of the first-strand cDNA product provides the annealing site for the 5′-amplimer during the PCR reaction. Homopolymer tailing catalysed by terminal transferase is controlled to allow the addition of 17 to 20 dG residues. Under the conditions of dGTP concentration, temperature and excess enzyme given in *Protocol 4* the tailing reaction is self-limited and gives a narrow distribution of tail sizes. Homopolymer tailing with dG is generally preferred to tailing with dC residues (5, 6).

Protocol 4. dG homopolymer tailing reaction

1. Resuspend the 'pellet' from *Protocol 3*, step 5 in $7\,\mu l$ H_2O.

2. Combine the following components, on ice, to give a final volume of $20\,\mu l$.

   - H_2O to $20\,\mu l$
   - First-strand cDNA $7\,\mu l$
   - $5 \times$ tailing buffer[a] $4\,\mu l$
   - 2 mM DTT $1\,\mu l$
   - 5 mM cobalt chloride $2\,\mu l$
   - $20\,\mu M$ dGTP $5\,\mu l$
   - Terminal transferase (Boehringer)   25 units

3. Incubate at $37\,°C$, 20 min.

4. Add $4\,\mu l$ 100 mM EDTA and $2\,\mu l$ 1 M NaCl to stop the reaction.

5. Add $1\,\mu l$ 10% (w/v) CTAB and spin in a microcentrifuge at $4\,°C$, 20 min, to precipitate the tailed first-strand cDNA.

Protocol 4. *Continued*

6. Carefully pipette off the supernatant and dissolve the 'pellet' in 10 μl 1 M NaCl.
7. Add 0.5 μl glycogen (0.5 mg/ml) as a carrier and 30 μl of cold ethanol. Precipitate overnight at −20°C.
8. Spin in a microcentrifuge at 4°C, 20 min. Wash the pellet in 70% ethanol and respin at 4°C, 10 min.

^a 5 × tailing buffer: 1 M potassium cacodylate, 125 mM Tris–HCl, pH 7.2.

2.5 RNA hydrolysis

Following homopolymer tailing the RNA which formed the template for first-strand cDNA synthesis is hydrolysed and removed from the reaction mix as described in *Protocol 5*.

Protocol 5. RNA hydrolysis

1. Resuspend the 'pellet' from *Protocol 4*, step 8 in 20 μl 50 mM NaOH, 2 mM EDTA and incubate at 65°C, 60 min.
2. Add 3 μl 3 M sodium acetate, pH 5.2, and 70 μl ethanol. Precipitate overnight at −20°C.
3. Spin in a microcentrifuge at 4°C, 20 min. Wash the 'pellet' in 70% ethanol, respin at 4°C, 10 min.
4. Resuspend in 20 μl H_2O

---

3. PCR amplification

In the example shown in *Figure 2* PCR amplification (see *Protocol 6*) of homopolymer-tailed first-strand cDNA utilized the following amplimers:

- A (oligo-dT$_{17}$-*Not*I); GCGGCCGCTTTTTTTTTTTTTTTTTTT
- B (oligo-dC$_{14}$-*Eco*RI); AAGGAATTCCCCCCCCCCCCCCC

During the first PCR cycle amplimer B anneals to the homopolymer-tailed 3'-end of the first-strand cDNA (see *Figure 1*). The annealing temperature for this step is 58°C to ensure specificity of primer:tail pairing.

In the second cycle the double-stranded cDNA is denatured allowing both A and B amplimers to anneal to their complementary priming sites. This step is carried out at a lower annealing temperature (40°C) to ensure good priming from the A:T rich 3'-end and results in two copies of each original cDNA. Subsequent cycles are performed at an annealing temperature of 58°C as the 3'-end has now incorporated the GC rich tail of amplimer A. If primer A is

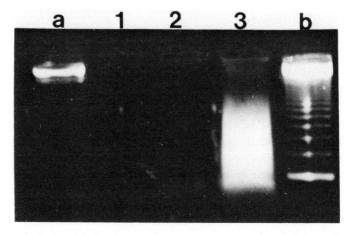

Figure 2. Analysis of first-strand cDNA synthesis. One nanogram of oligo-dG-tailed first-strand was amplified by PCR as described in *Protocol 6*. The products were separated on a 2% Nusieve GTG low-melting-point gel. **Lane a**, λ *Hind*III digested DNA; **lane b**, 123 bp markers; **lane 1**, PCR mix without amplimers; **lane 2**, PCR mix without cDNA; **lane 3**, PCR mix with first-strand cDNA and amplimers showing amplified products.

used to direct first-strand cDNA synthesis the low temperature annealing step is unnecessary. However, in our experience primer A does not prime first-strand cDNA synthesis with the same efficiency as oligo-dT.

Protocol 6. Second-strand synthesis and cDNA amplification

1. Combine the following components in 0.5 ml microcentrifuge tubes, on ice, to give a final reaction volume of 100 μl.
   - H₂O   to 100 μl
   - 10 × PCR buffer^a 10 μl
   - 10 × dNTP mix^b 10 μl
   - 100 pmol 5′ amplimer   1 μl
   - 100 pmol 3′ amplimer   1 μl
   - *Taq* polymerase   2 units
   - Oligo-(dG)-tailed cDNA   1–10 ng
2. Overlay the PCR reaction mix with ~100 μl mineral oil
3. Programme the primary amplification as follows:
   - denature the products at 96°C, 2 min,
   - 1 cycle (96°C, 2 min; 58°C, 2 min; 72°C, 3 min)
   - 1 cycle (96°C, 2 min, 40°C, 2 min, 72°C, 3 min)
   - 15 cycles (96°C, 2 min; 58°C, 2 min; 72°C, 3 min).

Protocol 6. *Continued*

4. Scale up reaction fourfold by dividing primary amplification products into four aliquots and add fresh primers, buffer, dNTP mix, *Taq* polymerase. Cycle 5 times (96°C, 2 min; 58°C, 2 min; 72°C, 3 min).

5. Combine the PCR products and ethanol precipitate, redissolve in 10 μl.

6. Fractionate the PCR products by electrophoresis through a 1–2% NuSieve-GTG LMP (FMC Bioproducts) gel against a 123 bp DNA ladder or other appropriate size marker (see *Figure 2*).

7. Excise fragments from gel into suitable size classes (e.g. see Section 3.2). Purify excised fragments by placing the gel slice in a Spin-X filter (Costar). Freeze then spin in a microcentrifuge, 15 min at r.t. (see Chapter 12, Section 2.3.2). Alternatively, purify according to the freeze–squeeze method (7) or use GeneClean (Bio101 Inc.).

8. Differentially, reamplify according to length by cycling 96°C, 2 min; 58°C, 2 min; 72°C, 3 min length as follows:

   - ~ 500 bp   1 cycle
   - 500 to 1500 bp   6 cycles
   - 1500 to 2000 bp   12 cycles
   - > 2000 bp   18 cycles

---

[a] 10 × PCR buffer: 100 mM Tris–HCl, pH 8.3, 500 mM KCl, 40 mM MgCl$_2$, 0.1% gelatin.
[b] 10 × dNTP mix; 5 mM dATP, dCTP, dGTP, dTTP.

3.1 Alternative amplification strategy

It is sometimes possible to further reduce the representation of small PCR products by modifying *Protocol 6* in the following manner. After 5 cycles of amplification (rather than 15) in step 3 carry out a size fractionation as described in steps 6 and 7. In this case PCR products will not be visible so recovery of the correct size fractions depends on the excision of regions of agarose defined by size markers. Amplify aliquots of the PCR products according to step 4 but for 20 to 25 cycles rather than 5. Analyse a sample of each PCR reaction by agarose gel electrophoresis to determine whether a second size fractionation is necessary. In our experience, however, the conditions in *Protocol 6* give good results and represent our preferred approach.

3.2 Size fractionation of primary PCR products

Four size fractions were isolated from the gel (*Figure 2*; *Protocol 6*, step 6) during the construction of the sample library and consist of fragments:

- <500 bp
- 500 to 1500 bp

- 1500 to 2000 bp
- >2000 bp

Samples of these primary amplification products were used for library construction as described in Section 4.2. To assess the relative size enrichment within the PCR products aliquots (approx. 10 ng) of each size fraction were electrophoresed through a 1% agarose gel and Southern blotted (4). The resulting filter was hybridized with a hexaprime-labelled probe prepared from a cDNA clone derived from a homologous gene (clone 2) that is known to be highly abundant and constitutively expressed. The transcript size of this gene is approximately 1150 bp and we observe the greatest enrichment of hybridization in the gel lane containing PCR products of 500–1500 bp as expected (see *Figure 3*).

3.3 Sequence representation within cDNA

To ensure that low-abundance transcript-derived cDNA had been amplified a further series of hybridization experiments were performed. To provide sufficient material for these analyses aliquots of primary PCR products were reamplified (see *Protocol 6*, step 8). Three size fractions,

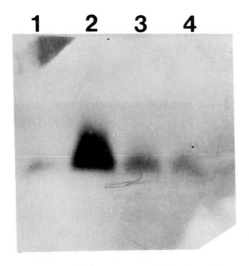

Figure 3. Southern blot analysis of initial PCR products. The PCR products from *Figure 1* were excised from the gel as four size fractions purified through Spin-X columns and were differentially amplified as shown in *Table 2*. 10 ng of each size selected class was separated on a 1% Nusieve gel; **lane 1**, <500 bp; **lane 2**, 500–1500 bp; **lane 3**, 1500–2000 bp; **lane 4**, >2000 bp. The gel was Southern blotted and the filter probed with hexaprime-labelled insert of clone 2, a highly abundant, constitutively expressed gene of transcript size 1150 bp. Strongest hybridization to the probe after stringent washing (65°C; 0.5 × SSC, 0.1% SDS) is seen in **lane 2** corresponding to the PCR products of size 500–1500 bp.

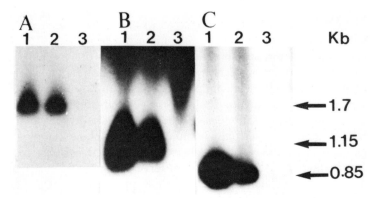

Figure 4. Southern blot analysis of primary PCR products. The products were divided into three size classes prior to secondary amplification (*Table 2*). Ten nanograms of the products were separated through 1% NuSieve low-melting-point gels loading **lane 1**, products >1500 bp; **lane 2**, products 500–1500 bp; **lane 3**, products <500 bp and probed with hexaprime-labelled low and high abundance genes of long and short transcript length. **Blot A**; hybridization to clone 3 reveals a 1.7 kb signal after 3 days' exposure to X-ray film; **Blot B**, hybridization to clone 2 reveals a 1.15 kb signal in **lanes 1** and **2** after 1 day exposure; **Blot C**, hybridization to clone 1 reveals a 0.85 kb signal in **lanes 1** and **2** after 10 days' exposure.

- >1500 bp (*lane 1, Figure 4*)
- 500 to 1500 bp (*lane 2, Figure 4*)
- <500 bp (*lane 3, Figure 4*)

derived from the material collected in Section 3.2 were reamplified by subjecting to the number of additional PCR cycles shown in *Protocol 6*. Aliquots of this reamplified material were Southern blotted and probed with three clones which gave the following results (*Figure 4*):

- CLONE 1   850 bp low to moderate abundance; blot C, *lanes 1* and *2* only.
- CLONE 2   1150 bp high abundance; blot B, *lanes 1* and *2* only.
- CLONE 3   1700 bp moderate abundance; blot A, predominantly in *lane 1*.

The appearance of each transcript approximately correlates with its abundance level since exposure times of the filters to X-ray film vary as indicated in *Figure 4* from 1 day for clone 2, 3 days for clone 3 to 10 days for clone 1.

4. Library construction

4.1 Choice of vector

Vectors of both bacteriophage and plasmid origin have been widely used for cDNA library construction and the considerations governing choice of a vector for a PCR-based library are the same as those for a traditional cDNA library.

To a large extent vector choice is a matter of personal preference and depends on the subsequent screening strategy and applications of isolated clones.

We prefer the bacteriophage vector λZAP which offers several advantages as the cloning vehicle.

(a) Insertional inactivation of the β-galactosidase gene in recombinant phage provides a rapid visual assessment of library size.

(b) λZAP can be rapidly converted to pBlueScript with no intervening *in vitro* subcloning steps.

(c) T3 and T7 promoters allow specific and rapid generation of RNA probes for *in situ* hybridization studies.

(d) The library can be antibody screened.

4.2 Host cells

λZAP is handled in the same manner as other λ bacteriophage and is generally plated on either of two host strains supplied by Stratagene. These strains, BB4 and XL1-Blue, may be used interchangeably although the former does grow significantly more rapidly and gives higher plaque titres. However, we prefer to use XL1-Blue since this strain is *recA⁻* compared with BB4 which is *recA⁺*. XL1-Blue often provides higher yields of DNA. Plating cells suitable for infection by λZAP are easily prepared according to *Protocol 7*.

Protocol 7. Preparation of plating cells

1. Streak out XL1-Blue on a fresh $2 \times TY^a$ plate containing $12.5\,\mu g/ml$ tetracycline and grow overnight.

2. Pick an isolated colony into $100\,ml$ $2 \times TY$ containing $10\,mM$ $Mg\,SO_4$, $10\,mM$ maltose and grow to an $OD_{600} = 0.5$.

3. Centrifuge ($\sim 3000\,g$) for 5 min, decant supernatant.

4. Resuspend cells initially in $20\,ml$ $10\,mM$ $MgSO_4$ then adjust vol. to give a suspension of $OD_{600} = 1$. Freshly prepared cells work best although they can be stored for up to a week at 4°C.

^{a} $2 \times TY$: $16\,g$ bactotryptone, $10\,g$ yeast extract, $5\,g$ NaCl per litre, and adjust pH to 7.4 with NaOH.

4.3 Digestion of DNA

4.3.1 Vector DNA

Vector DNA should be highly purified, and we favour caesium chloride banding of the bacteriophage followed by repeated phenol extractions as described by Davis (4). The vector is then prepared according to *Protocol 8* by digesting with restriction enzymes yielding ends complementary to those of the PCR products.

4.3.2 PCR products

Digest the PCR products with the appropriate restriction enzymes (in this case *Not*I and *Eco*RI). Digestion of the PCR product has the added benefit of removing any extraneous residues (usually A) that might be added to the DNA during PCR. The size-fractionated PCR products from *Protocol 6*, step 7 were used to construct the library. These primary PCR products were not reamplified to prevent the differential amplification of higher abundance sequences. The selective reamplification procedure (*Protocol 6*, step 8, and Section 3.2) was only used to assess the representation and abundance of products derived from transcripts.

Protocol 8. Preparation of vector

A. *Vector DNA*

1. Combine the following components in a 1.5 ml microcentrifuge tube
   - H_2O to 10 µl
   - λZAP   1 to 5 µg
   - *Not*I   2 units
   - *Eco*RI   2 units
   - 10 × buffer[a]   6 µl

2. Digest for 1 h at 37°C.

3. Add 1 unit calf intestinal phosphatase (CIP; Boehringer-Mannheim) to the linearized vector and incubate for 30 min at 37°C to dephosphorylate (4).

4. Extract once with TE-saturated phenol (4), once with chloroform and precipitate with 0.1 vol. 3M sodium acetate, pH 7, and 2.5 vol. ethanol, overnight at −20°C or 30 min at −70°C.

5. Redissolve pellet in 2 to 10 µl TE to give a concentration of ~500 ng/µl.

B. *PCR products*

6. Combine the following components in a 1.5 ml microcentrifuge tube
   - H_2O to 10 µl
   - PCR products   200 ng
   - *Not*I   0.5 units
   - *Eco*RI   0.5 units
   - 10 × buffer[a]   2 µl

7. Digest for 1 h at 37°C.

^a 10 × buffer: Pharmacia 'One-Phor-All' with Triton ×-100 added to 0.01%. *Not*I requires a final concentration of 2 × 'One-Phor-All' buffer. *Eco*RI digests well under these conditions.
^b TE: 10mM Tris–HCl, pH 7.5, 1 mM EDTA.

4.4 Ligation and packaging

Suitably digested vector and PCR amplified DNA are ligated under conditions that generate concatameric recombinant molecules suitable as substrates for the packaging reaction into lambda particles. The procedure for these steps is given in *Protocol 9*. The quality of the final library is obviously dependent on the success of the ligation and of the packaging reactions. The larger the number of p.f.u. generated the greater the chance of obtaining a target clone. We purchase suitable lambda packaging mixes and have had excellent results with Gigapack Plus (Stratagene).

Protocol 9. Ligation and packaging reactions

A. *Ligation*

Set up ligations in a final volume of 5 μl. Various molar ratios of vector:insert. may be tested and two vector only ligations containing (a) digested vector, and (b) digested plus CIP-treated vector may be included to assess the efficiency of dephosphorylation and ligation.

1. Combine the following components:
- H₂O — to 5 μl
- 5 × Ligase buffer^a — 1 μl
- 50 mM ATP — 1 μl
- Vector DNA (from *Protocol 8*, part A) — 1 μg
- Insert (from *Protocol 8*, part B) — 40 ng
- T4 DNA ligase — 1 unit

2. Incubate for 1 h at r.t. then overnight at 4°C.

B. *Packaging*

Use a suitable packaging system such as Gigapack Plus (Stratagene) according to the manufacturers instructions.

3. Add 1 μl of ligation mix from step 2 to 10 μl of freeze–thaw lysate as it is thawing.^b

4. Place the tube on ice. Rapidly thaw the sonic extract and add 15 μl to the freeze–thaw tube.^b Leave at 22°C for 2 h.

5. Add 500 μl SM buffer^c and 20 μl chloroform, mix gently, spin for a few seconds to precipitate debris, then store at 4°C.

Protocol 9. *Continued*

6. Dilute a 10 μl aliquot of the supernatant to 10^{-6}, 10^{-7}, and 10^{-8} in SM then transfect freshly prepared plating cells (see *Protocol 7*) of a suitable host strain (e.g. BB4 or XL1-Blue) to titre.

7. Mix 1 μl of diluted phage with 200 μl plating cells and incubate in a 37°C water bath for 15 min.

8. Mix 3 ml molten top agarose (TY plus 0.7% agarose) at 48°C with 50 μl 0.5 M X-Gal in dimethylformamide and methanol (1:1 v/v) and 50 μl 250 mg/ml IPTG. Add the infected cells and pour on to a pre-warmed (37°C) LB plate. Allow to set then invert and incubate overnight at 37°C.

[a] 5 × Ligase buffer: 200 mM Tris–HCl, pH 7.6, 50 mM $MgCl_2$, 5 mM DTT.
[b] The freeze–thaw lysate tube should not be allowed to thaw until the contents of the sonic extract tube plus DNA are being transferred into it. Also take care not to introduce air bubbles into the packaging reaction during the addition of DNA to the sonic extract or its transfer to the freeze–thaw lysate tube. Both these factors reduce packaging efficiency.
[c] SM buffer: 0.1 M NaCl, 10 mM $MgSO_4$, 50 mM 1M Tris–HCl, pH 7.5, 0.01% gelatin per litre.

4.5 Assessment of library quality

The quality of the library can be assessed by the proportion of recombinants compared with wild-type phage. We use two parameters to measure this, (a) a simple colour selection that provides a preliminary screen, and (b) a rapid PCR screen that provides an indication of the average cDNA insert size.

4.5.1 Colour selection

The IPTG and X-Gal selection step in *Protocol 9* provides an indication of the relative numbers of recombinant (white) versus wild-type (blue) phage. We only use this colour reaction during this initial titration since the level of blue colouration is very poor compared with the similar M13 selection system. A newer version of the vector, λZAP II is reported to have a much more intense colour selection (Stratagene).

4.5.2 PCR-based plaque screen

A highly versatile procedure described by Gussöw and Clackson can be used to amplify DNA directly from either bacterial colonies or plaques (8; see also Chapter 12). We use such a screen described in *Protocol 10*, to amplify insert DNA from a random selection of plaques from the titration plate. The primers used for amplification are complementary to the *lacZ* sequences flanking the multiple cloning site of λZAP and many other vectors such as pBlueScript, the pUC series and M13 series. Forward and reverse M13 sequencing primers or T3 and T7 primers are perfectly adequate for this step.

Protocol 10. Rapid PCR screening of plaques

1. Prepare a PCR pre-mix containing the following components per 20 μl per plaque to be screened;
   - H_2O to 20 μl
   - 10 × PCR buffer[a]   1 μl
   - 2.5 mM dNTPs   2 μl
   - Primer 1   10 pmol
   - Primer 2   10 pmol
   - *Taq* polymerase (AmpliTaq, Cetus)   2.5 units

2. Dispense 20 μl aliquots of PCR pre-mix to 0.4 ml microcentrifuge tubes and toothpick a random plaque into each aliquot.

3. Overlay with 20 μl mineral oil and amplify for 30 cycles of [94°C, 1 min, 55°C, 1 min, 72°C, 2 min] then hold for 5 min at 55°C. Cool and hold at 4°C.

4. Electrophorese an aliquot of each reaction (2 μl is usually sufficient) on a 2% NuSieve GTG LMP gel (FMC Bioproducts) containing 0.5 μg/ml ethidium bromide. Visualize under UV illumination (see *Figure 5*).

[a] 10 × PCR buffer: 100 mM Tris–HCl, pH 8.3, 500 mM KCl, 15 mM $MgCl_2$, 0.1% gelatin.

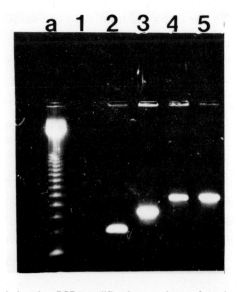

Figure 5. Agarose gel showing PCR amplification products of randomly selected plaques from the λZAP library. **Lane a**, 123 bp marker; **lane 1**, PCR with amplimers only; **lanes 2** to **5**, PCR of randomly picked plaques.

Transfection of host cells with the packaged ligation mix as described in *Protocol 9* generated a library of 3.5×10^5 clones of which approximately 96% were recombinants as assessed by the X-Gal/IPTG colour screen. Several white plaques were picked and characterized by PCR screening according to *Protocol 10*. All contained inserts ranging in size from 230 to 900 bp and with an average size of 470 bp. Four such clones are shown in *Figure 5* together with no plaque (primers only) control (*lane 1*).

5. Library screening

Identification of clones corresponding to differentially expressed genes can be achieved by parallel hybridization screening of duplicate plaque-lift filters. If possible, screen two filters per plate with each probe (i.e. lift four filters from each plate). In the example cited we have used radiolabelled cDNA probes prepared from RNA isolated from

(a) infected material from which the library was constructed, and

(b) healthy control tissue.

This screen should allow identification of clones representing genes either up-regulated or down-regulated following infection.

5.1 Probe preparation

The probes used to differentially screen the PCR-based cDNA library are synthesized according to *Protocol 11* from 10 µg of total RNA prepared from the material used to construct the library and from healthy tissue.

Protocol 11. Preparation of differential cDNA probes

1. Denature 10 µg total RNA at 70°C for 5 min and transfer to ice.
2. Combine the following components, in order, to give a final vol. of 50 µl.
   - H_2O to 50 µl
   - 5 × AMV-RT buffer (see *Protocol 2*)   10 µl
   - Oligo-dT (100 mg/ml)   5 µl
   - 2 mM each dATP   2.5 µl
   - 10 mM spermidine–HCl   2.5 µl
   - 80 mM sodium pyrophosphate   1 µl
   - HPRNI   1 µl
   - Total RNA   10 µg
   - $[\alpha-{}^{32}P]$ dCTP   50 µCi
   - AMV reverse transcriptase   10–20 units

3. Incubate at 42°C, 15 min.

4. Add 5 µl chase solution (10 mM each dATP, dCTP, dGTP, and dTTP) and incubate 20 min at 42°C.

5. Stop reaction by adding 50 µl 0.3 M NaOH, 30 mM EDTA and boil for 5 min.

6. Add 15 µl 1 M Tris–HCl, pH8, 50 µl Tris-saturated phenol, vortex and spin in a microcentrifuge for 1 min. Add 50 µl 4 M ammonium acetate and 200 µl absolute ethanol to supernatant and precipitate overnight at −20°C.

7. Spin in a microcentrifuge and wash pellet sequentially as described in *Protocol 2*. Resuspend in 200 µl TE (see *Protocol 8*).

8. Determine the amount of radioactivity in each probe by spotting 1 µl on activated DE-81 filters and count in scintillation fluid (^{32}PP channel) as described by Berger (9). Adjust the volumes in step 7 to give equivalent c.p.m./µg of each probe. Assuming a maximum of 200 ng of the 10 µg total RNA is used in first-strand synthesis, probes of approximately 10^7 c.p.m./µg can be achieved.

9. Denature probe by boiling for 5 min, snap cool on ice and add to the hybridization fluid.

5.2 Preparation of plaque-lift filters

Plate sufficient p.f.u. to ensure that clones derived from low abundance transcripts are represented. On the assumption that a low abundance message is represented once in 25 000 transcripts for a typical higher eukaryotic cell we recommend plating around 5×10^4 to 10^5 p.f.u.

Protocol 12. Plaque lifts

1. Plate PCR-based cDNA library to give a density of 2000 plaques per 9 cm^2 plate and incubate 8–10 h at 37°C.

2. Chill plates at 4°C for a minimum of 2 h prior to use.

3. Using a soft pencil, label pairs of pre-cut filters with numbers, and either A or B.

4. Lift plaques on to pre-cut Hybond-N nylon membrane (Amersham) allowing 1 min adsorption for the first lift and 5 min for subsequent replica lifts. Mark orientation of first set of filters by needle-piercing nylon membrane and agar through to base of Petri dish and mark with fine marker pen. Mark subsequent filter lifts by piercing membrane with needle aligned with marker on base of Petri dish.

5. Remove filters and air-dry, plaque-side uppermost.

Protocol 12. *Continued*

6. Float, plaque-side upwards, in denaturing solution (1 M NaCl, 0.5 M NaOH) for 20 sec, followed by 20 sec in neutralizing solution (0.5M Tris–HCl, pH 7.4, 1.5 M NaCl).

7. Wash filters in 3 × SSC*ᵃ* for 15 min with gentle shaking, and air-dry for 30 min.

8. Bake at 80°C for 2 h.

ᵃ 20× SSC: 3 M NaCl, 0.3 M Na₃ citrate, pH 7.

5.3 Hybridization screening

The filters (prepared in *Protocol 12*) are initially pre-hybridized to block non-specific binding sites before hybridizing to the appropriate probes (see *Protocol 11*) as described in *Protocol 13*.

Protocol 13. Pre-hybridization, hybridization, and washing of filters

A. *Pre-hybridization*

1. Place the complete set of A replica filters in one plastic box and the B set in a second box. Pre-wet filters by rinsing in 3 × SSC.

2. Add 3–4 ml pre-hybridization fluid*ᵃ* per filter.

3. Pre-hybridize the filters at 65°C for a minimum of 6 h.

B. *Hybridization*

4. Denature the probes prepared in *Protocol 11* by boiling for 5 min. and rapidly cool in ice.

5. Remove 5 ml pre-hybridization solution and mix with the appropriate probe.

6. Return this solution to the appropriate box, mix well, and hybridize at 65°C for 24 h.

C. *Washing*

7. Following hybridization, wash filters at 65°C sequentially in two changes of

  • 5 × SSC, 0.1% SDS then

  • 3 × SSC, 0.1% SDS then

  • 2 × SSC, 0.1% SDS

  for 30 min per wash.

8. Lay filters in matched pairs aligning orientation marks highlighted with radioactive ink. Cover with Saran wrap and expose to pre-flashed X-ray film at $-70\,°C$, using intensifier screens, for 3–10 days

---

a Pre-hybridization solution contains $5 \times$ SSPE ($20 \times$ SSPE $= 3\,M$ NaCl, $0.2\,NaH_2PO_4.H_2O$, $0.02\,M\,Na_2EDTA$, adjust pH to 7.4 with $10\,M$ NaOH), 6% PEG 6000, 0.5% Marvel (skimmed milk powder), 1% SDS, 0.1% sodium pyrophosphate, $250\,\mu g/ml$ ultrasonicated herring sperm DNA (denature by boiling for 10 min prior to use). All solutions are sterilized by autoclaving, except Marvel which is either heated in a microwave (600 to 700 W) for 2 min or autoclaved at only 5 p.s.i. for 5 min. It is essential that the milk solution does not turn brown.

5.4 Secondary and tertiary screening

To ensure purification of single clones, putative positive plaques (see *Figure 6*) are screened through two further rounds of differential hybridization as follows. Pick areas of top agar around the putative positive plaques into $500\,\mu l$ SM (see *Protocol 9*), add $20\,\mu l$ chloroform then vortex for 10 sec before storing at $4\,°C$. Prepare a dilution series in SM to 10^{-5} and plate $10\,\mu l$ aliquots after incubation with plating cells as described in *Protocol 9*. Lift two to four replica filters from each plate containing 500 to 1000 plaques (see *Protocol 12*) and differentially screen replica filters with each probe as described in *Protocol 13*.

A further round of screening of single plaques as above is usually necessary to ensure single plaques have been purified.

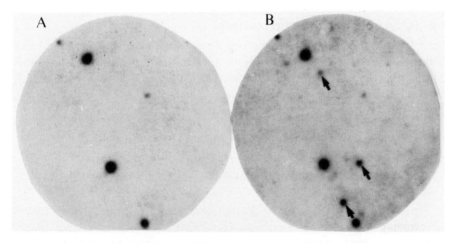

Figure 6. Differential hybridization of ^{32}P-labelled first-strand cDNA probes to a sample of the cDNA library. **A** and **B** are replica filters lifted from a single plate containing 2000 plaques. cDNA probes were synthesized from $10\,\mu g$ of total RNA from the appropriate tissue; **filter A**, healthy tissue; **filter B**, infected tissue. Arrows denote three plaques isolated for further analysis. The large spots represent clones of highly expressed gene detected by both probes.

6. Northern blot analysis

It is essential to confirm that positive clones represent differentially expressed transcripts by Northern blot analysis to appropriate RNA samples. In this example it should hybridize to infected but not to healthy tissue.

6.1 Preparation of hexaprime-labelled probe

The insert from a putative positive clone can be rapidly amplified and labelled according to *Protocol 15* for use as a probe in Northern analysis.

Protocol 14. Preparation of hexaprime-labelled probe

1. Pick plaque of interest into 20 µl SM buffer and vortex.
2. Use 10 µl of supernatant in a 20 µl PCR reaction as in *Protocol 10*.
3. Electrophorese PCR products through a 2% NuSieve GTG LMP gel including 0.5 µg/ml ethidium bromide.
4. Excise band into a Spin-X filter, freeze, and spin in a microcentrifuge for 20 min at r.t. (see *Protocol 6*, step 7).
5. Hexaprime label (10) 25 ng purified fragment for 5 h at r.t.
6. Remove unincorporated isotope by passage through a G-50 Sephadex column (11).

6.2 RNA gel and hybridization

Total RNA was prepared from various tissue samples including infected and healthy tissue. *Protocol 15* describes the gel fractionation blotting and hybridization with the probes prepared in Section 6.1.

Protocol 15. Preparation of RNA gel

1. Electrophorese 5–10 µg total RNA through 0.6 M formaldehyde agarose gels containing 0.5 µg/ml ethidium bromide (4).
2. Photograph gel together with a ruler to allow subsequent assignment of transcript size.
3. Rinse gel in 10 × SSC, 2 × 20 min changes at 42°C to remove excess formaldehyde.
4. Blot on to Hybond-N nylon membranes (Amersham) and pre-hybridize for at least 4 h in 50% formamide, 5 × SSC, 1 × Denhardts solution (11), 250 mM $Na_4P_2O_7$ (pH 6.5), 100 µg/ml sonicated herring sperm DNA.
5. Heat denature the probe (from *Protocol 14*) by boiling for 10 min then snap cool on ice.

6. Combine the denatured probe with 5 ml hybridization fluid (see step 4) and hybridize 24 h at 42°C.

7. Wash filter sequentially in two changes each of 5 × SSC, 0.1% SDS, then 0.2 × SSC, 0.1% SDS at 42°C for 30 min per wash. Wash 2 × in 0.2 × SSC, 0.1% SDS at 65°C for 30 min per wash.[a]

8. Expose to X-ray film at −70°C, with intensifier screens.

[a] It is advisable to expose the filter to X-ray film before the 65°C wash to provide a film for comparison with image following the more stringent wash. The less stringent wash often also allows visualization of the rRNA bands which can be useful for transcript size determination.

Figure 7 shows a Northern blot of a putative positive clone, prepared and selected according to the procedures described in this chapter, showing hybridization to infected (*lane 4*) but not healthy tissue probes (*lanes 1 to 3*). The figure also shows hybridization of a clone representing a constitutively expressed gene emphasizing the differential hybridization of the selected clone.

7. Conclusion

The approach outlined in this chapter should be generally applicable to any biological system where the amount or proportion of differentiated tissue is

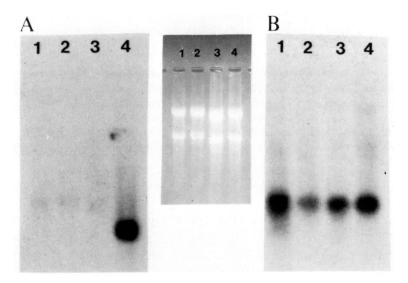

Figure 7. Northern blot analysis. RNA was isolated from root material following various treatments. The probe was prepared from the insert of a clone selected through three rounds of differential screening as being infection-specific.

severely limited. Representative cDNA libraries can be constructed rapidly and while the majority of clones are not full-length we have successfully identified and characterized a number of interesting cDNAs which have subsequently been used to isolate corresponding genomic clones.

Acknowledgements

This work was supported in part by grants from Enichem Americas, AFRC, and SERC. S.J.G. is supported by a Royal Society University Research Fellowship. We acknowledge Kim Hammond for providing clones 1–3.

References

1. Gurr, S. J., McPherson, M. J., Scollan, C., Atkinson, H. J., and Bowles, D. J. (1991). *Molecular and General Genetics*. **226,** 361.
2. Belyavsky, A., Vinogrdora, T., and Rajewsky, K. (1989). *Nucleic Acids Research*, **17,** 8.
3. Gurr, S. J. & McPherson, M. J. (1992). In *Molecular plant pathology: A practical approach* Vol. 1 (ed. S. J. Gurr, M. J. McPherson, and D. J. Bowles), p. 109. IRL Press at Oxford University Press, Oxford.
4. Davis, L. G., Dibner, M. D., and Battey, J. F. (1986). *Basic methods in molecular biology*. Elsevier Science Publishers, New York.
5. Dugaiczyk, A., Robberson, D. L., and Ullrich, A. (1980). *Biochemistry*, **19,** 5869.
6. Otsuka, A. (1981). *Gene*, **13,** 339.
7. Horton, R. M. and Pease, L. R. (1991). In *Directed mutagenesis: A practical approach* (ed. M. J. McPherson), p. 217. IRL Press at Oxford University Press, Oxford.
8. Gussow, D. and Clackson, T. (1989). *Nucleic Acids Research*, **17,** 4000.
9. Berger, S. L. (1987). In *Methods in enzymology*, Vol. 152 (ed. S. L. Berger and A. R. Kimmel), p. 49. Academic Press, New York and London.
10. Feinberg, A. P. and Vogelstein, B. (1983) *Nucleic Acids Research*, **14,** 2229.
11. Sambrook, J., Fritsch, E. F., and Maniatis, T. (1989). *Molecular cloning: A laboratory manual*, 2nd edn. Cold Spring Harbor Laboratory Press, Cold Spring Harbor, NY.

11

PCR with highly degenerate primers

MICHAEL J. McPHERSON, KERRIE M. JONES, and SARAH JANE GURR

1. Introduction

Primers for use in the polymerase chain reaction are usually unique oligonucleotides designed from a known DNA sequence. However, for an uncloned gene, where obviously no DNA sequence data are available, the PCR may be used in cloning strategies by designing highly degenerate primers from protein sequence data (1–3). Such degenerated primers have been used successfully to isolate members of multigene families and genes from diverse species. The amplified DNA represents part of the gene of interest and can therefore be used as a *homologous* hybridization probe, allowing high stringency screening of gene libraries to isolate corresponding cDNA or genomic clones. The PCR amplified DNA may also be used to probe Southern or Northern blots.

Highly degenerate primers, based on a single region of peptide sequence data, may also be used in rapid amplification of cDNA ends (RACE) PCR (4). This allows the efficient amplification of the 5'- and 3'-ends of a target cDNA and, if required, the subsequent generation of a full-length cDNA by virtue of the overlapping sequences between the two original primers.

This chapter outlines the use of highly degenerate oligonucleotide primers, designed from peptide sequence data, to amplify corresponding DNA sequences. In particular two approaches are considered

(a) the use of sequence derived from a single protein to isolate the corresponding gene, and

(b) the use of multiple sequence alignments to generate 'universal' primers capable of amplifying corresponding regions of the same gene from diverse species.

This latter approach is particularly powerful when coupled with a nested PCR strategy.

2. Source of template DNA

Either genomic DNA or cDNA are suitable for amplification by the PCR with highly degenerate primers.

2.1 Genomic DNA

Various methods have been used to prepare genomic DNA from bacteria, fungi, plant, and animal tissues. It may be necessary to purify plant DNA by CsCl density gradient centrifugation to remove cell wall components; bacterial, fungal, and animal DNAs do not usually require this step. Dissolve the DNA in double distilled water at a concentration of around 0.1 mg/ml.

2.1.1 Bacteria

DNA from both Gram negative and Gram positive strains can be prepared according the method of Marmur (5). Usually it is sufficient to prepare DNA on a small-scale according to a modification of this method (6).

2.1.2 Filamentous fungi

The spermidine buffer method of Azevedo *et al.* (7) is highly efficient and rapid. It will probably be necessary to perform five to seven phenol extraction steps.

2.1.3 Plants

DNA from a range of plants including tomato, potato, soybean, and barley have been prepared according to the method of Jofuku and Goldberg (8).

2.1.4 Animal tissue

DNA is prepared from animal tissue by a proteinase K-SDS procedure such as that described in ref. 9.

2.2 cDNA

Various methods for the isolation of total RNA and for first-strand cDNA synthesis from total RNA are described in Chapters 3, 4, 5, 10, 12, and 13 of this volume.

2.3 Contamination problems

The ability of the PCR to amplify minute amounts of template has the disadvantage that small quantities of contaminating DNA may generate confusing and misleading results where a gene is being amplified from a range of species. Chapters 2, 3, and 12 in this volume discuss problems of contamination in the PCR and introduce various strategies for the prevention of such problems. In addition to adopting of such prevention measures it is important that control reactions are performed in parallel with the test samples to detect any contamination problems. At least two controls are required, a 'no DNA' reaction and a 'no primers' reaction.

3. Highly degenerate PCR primers

3.1 Primers based on peptide sequence from a single protein

If a protein of interest has been identified, it is often possible to purify sufficient material to allow the determination of a limited amount of amino acid sequence data. Ideally, different regions of the protein should be sequenced to provide information for the design of at least two primers. Usually this is achieved by the proteolytic cleavage of the protein (by for example protease V8) to yield a limited set of peptide products for sequencing. Methods for the preparation, purification, and sequencing of such peptides are covered in a separate volume in the *Practical Approach* series (10). Primers are designed to represent all possible coding sequences for the selected region of peptide (see Section 3.3).

Sequence data from a single protein may actually provide sufficient data to facilitate the cloning of the same gene from a wide range of organisms if the gene is highly conserved. For example, it has been possible to amplify corresponding segments of plant annexin genes from tomato, arabidopsis, potato, and barley by using primers designed from tomato annexin peptide sequence data (M. F. Smallwood, S. J. G., M. J. M., K. Roberts, and D. J. Bowles, *Biochemical Journal*, **281,** 501 (see *Figure 1*). In cases where sequence data from a single protein does not prove sufficient for the selection of a gene family from a range of species it may be possible to use sequence alignment data (Section 3.2).

3.2 Primers based on multiple sequence alignments

When two or more complete or partial protein or DNA sequences are available from different organisms it is often possible to identify highly conserved regions of peptide sequence. Primers can then be designed from these regions by including base redundancies representative of all of the known peptide sequences (see Section 3.3). In other cases it may be sufficient to incorporate redundancies based only on a select subsets of the sequences.

We have used such an approach to amplify regions of glutamate dehydrogenase genes from prokaryotic and eukaryotic organisms. In this case the available sequence data were based primarily on bacterial and fungal sequences. A number of vertebrate GDH sequences were also available but only one plant-like sequence, from Chlorella (11). Since the intention was to study plant genes the contributions of the vertebrate sequences in the design of the oligonucleotides was reduced, though not totally excluded.

3.3 Design of highly degenerate primers

Back-translation from peptide sequence to the corresponding DNA coding sequence is complicated by the fact that most amino acids are encoded by more than one codon. In certain cases with organisms that display significant bias in codon usage, it is possible to design PCR primers based on this preferential codon usage. More often, however, primers must be designed to reflect the full codon redundancy in order to represent all possible peptide coding sequences (see *Figures 1* and *2*).

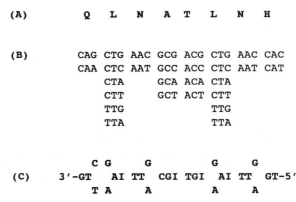

Figure 1. Design of highly degenerate oligonucleotides from peptide sequence data. (**A**) A segment of peptide sequence data from a tomato annexin shown in single letter code, (**B**) 'back-translation' of this peptide into all possible corresponding DNA codons, and (**C**) highly degenerate oligonucleotides designed from these data; note that the primer sequence in this example is complementary to the peptide coding sequence. Inosine has been incorporated at positions of fourfold redundancy. The potential number of sequences to which this oligonucleotide could hybridize is 8092. This primer was used with a second degenerate primer mix to generate the results shown in *Figure 6*. The presence of residues such as serine, arginine, and leucine that have many possible codons is not a limitation to the design and successful utilization of such primers.

Since some amino acids have only one or two possible codons (M, W, C, D, E, F, H, K, N, Q, Y) it is best to select regions rich in such residues for primer design. Other amino acids have up to six possible codons (L, R, S) and are best avoided. Unfortunately these simple selection rules cannot always be followed; the peptide sequence may be composed entirely of amino acids encoded by more than two codons. Unlike oligonucleotide probing experiments, where the degeneracy of the probe is a critical factor in determining the success of the experiment, PCR primers of very high degeneracy can be used successfully.

3.3.1 Mixed-base synthesis and inosine

Positions of four-fold base degeneracy may be incorporated into the oligo by 'mixed-base' synthesis; i.e. addition of all four nucleotide monomers at the coupling step resulting theoretically in 25% of the oligos with dA, 25% dG, 25% dC, and 25% dT at the degenerate position. For the examples shown in

(A)

```
R 112     P   C/Y   K    G    G   I/L/M  R

       5' CCI TGT AAA GGI GGI ATA AG 3'    X 12, 288
          AC   G             C G C
                             T C
                             T
```

```
K132      G    F    E    Q   T/I/A Y/F  K

       5' GGI TTT GAA CAA ATI IAT AA 3'    X 2, 048
          C    G      G GC  TC
```

```
D181      D    V    P    A   G/P D/N  I/M

       3' CTA CAI GGI CGI GGI TTA TA 5'    X 8, 192
          G              CC  C G
```

```
A232      A    T    G   Y/R  G  L/A/S F/V/L    X 73, 728

       3' CGA TGI CCI ATA CCI IGT GA
          G            G    A   A
          C                 C   C
          T
```

(B)

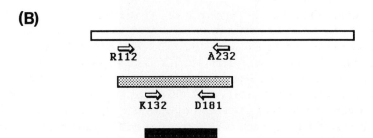

Figure 2. Design of highly degenerate primers from multiple sequence alignment data. **(A)** 'Universal' glutamate dehydrogenase primers designed from regions of peptide sequence shown as single letter code. The peptide sequence was derived from multiple sequence alignment of GDH sequences. Design principles are discussed in the text and the legend to *Figure 1*. **(B)** schematic outline of the positions of the primers from **(A)** within the GDH coding region. Initial amplifications were performed with R112 and A232 and nested amplifications with K132 and D181.

Figures 1 and *2* this would mean that mixtures of oligonucleotides representing all the possible degenerate sequences would be co-synthesized. In the most extreme example, primer D in *Figure 2*, the mixture would consist of 73 728 different oligonucleotide sequences. A simplification of this complex mixture can be achieved by incorporating the universal base inosine at positions of three- or four-fold base redundancy (3). For primer D (*Figure 2*) this reduces the complexity of the mixture from 73 728 to 72 different sequences.

3.3.2 3′-end of the primer

The three bases at the 3′-end of the primer *must* be perfectly matched with the template and should not include inosine (12). Beyond this '3′-anchor' even mismatched bases can be tolerated. For example, *Figure 3* (*lanes b* and *d*) shows PCR products generated in reactions where one of the primers contains two base mismatches at positions four and five from the 3′-end of the oligonucleotide, i.e. immediately adjacent to the three 3′-anchor bases. Selection of the 3′ bases is straightforward from peptide sequence data for a single protein. In the case of multiple alignment data however it is most convenient if the 3′-end of the primer corresponds to a totally conserved residue as in the examples shown in *Figure 2*.

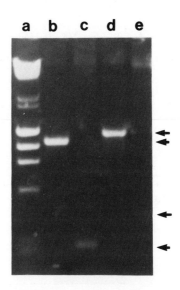

Figure 3. Amplification of bacterial genomic DNA sequences. 2% NuSieve GTG agarose gel stained with ethidium bromide. (**lanes b** to **e**) amplification of 1.1 (128 and 64), 0.3 (128 and 2048), 1.3 (256 and 64) and 0.45 (256 and 2048) kb respectively, generated by the PCR from *Clostridium symbiosum* genomic DNA. The figures in brackets indicate the degeneracy of the primers used to generate the product. Reactions containing 0.5 μg genomic DNA and 100 pmol each of the appropriate primers were performed according to *Protocol 1* for 30 cycles. One-tenth of each reaction was loaded.

3.3.3 5′-end of the primer

As with other PCR strategies, the 5′-end of the primer may be designed to add useful sequence features on to the PCR products. For example, if the fragments are to be cloned it is useful to include suitable restriction enzyme recognition sequences within the primers.

3.4 Nested primers

Often there may be sufficient peptide sequence data or regions of homology to allow the design of nested primers. In nested PCR a reaction is performed with the outer primer pair (R112 and A232; see *Figure 2*). An aliquot of this reaction is then reamplified with the inner pair of primers (K132 and D181). This approach significantly increases the sensitivity of the PCR, since two pairs of primers are required to amplify the target sequence for a final product to be generated. The GDH primers shown in *Figure 2* produced the results shown in *Figure 4*. *Figure 4b* shows the nested PCR amplification of unique bands from primary amplification products that contain several bands as shown in *Figure 4a*. Interestingly, the nested PCR demonstrates the presence of an intron of around 120 bp in the fungal DNA (*lane b*) compared with the bacterial (*lanes a* and *c*) samples. The presence of the intron was confirmed by DNA sequence analysis.

4. PCR amplifications using highly degenerate primers

Highly degenerate primers can be used to amplify DNA sequences, starting from either genomic DNA or cDNA. This section describes methods used for amplification from genomic DNA and the analysis of products. We have successfully used PCR primers ranging in potential redundancy from 64- to 73 728-fold, to amplify genomic DNA from species of bacteria, fungi, plants and animals. *Figures 3* and *5a* show amplifications from bacterial and fungal DNA respectively. *Figure 6* (*lane a*) shows the unique amplification of a fragment from tomato genomic DNA corresponding to a segment of an annexin gene. *Figure 4* shows amplifications from a range of organisms using the universal GDH primers shown in *Figure 2*. The PCR experiments were performed according to *Protocol 1*.

Protocol 1. PCR amplification of genomic DNA

1. For a single 50 μl PCR reaction add the following components to a 0.5 ml sterile Eppendorf tube:

- H$_2$O   to 50 μl
- 10 × PCR buffer[a]   5 μl

PCR *with highly degenerate primers*

A

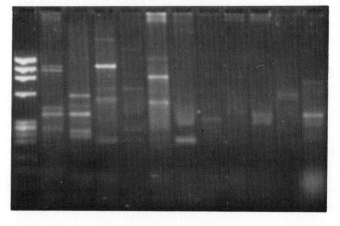

B

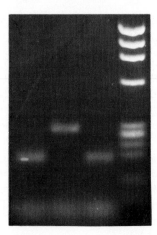

Figure 4. Universal primers and nested PCR. (**A**) 2% NuSieve gel showing the primary amplification products from a range of genomic DNA samples (**lanes b** to **k**). Primers R112 and A232 (see *Figure 2*) were used in PCR reactions according to *Protocol 1* for 35 cycles. (**lane a**) λ *Hind*III and φX174 *Hae*III size markers, (**lane b**) *Escherichia coli* (Gram negative bacterium), (**lane c**) *C. symbiosum* (Gram positive bacterium), (**lane d**) *Dactylium dendroides* (fungus), (**lane e**) *Physcomitrella patens* (moss), (**lane f**) potato, (**lane g**) tomato, (**lane h**) soy bean, (**lane i**) barley, (**lane j**) wheat, (**lane k**) Arabidopsis and (**lane l**) an unidentified cruciform. One-fifth of the total reaction products were loaded.

(**B**) Nested PCR amplification of reaction products shown in **lanes b, c,** and **d** of part (**A**). One-fiftieth of the initial PCR reaction was used as the template for nested PCR with the primers K132 and D181 (see *Figure 2*) according to *Protocol 1* (20 cycles). (**lane m**) λ *Hind*III and φX174 *Hae*III size markers, (**lane c'**) *C. symbiosum*, (**lane d'**) *D. dendroides* and (**lane b'**) *E. coli*. In each case a new and unique product can be seen. Products were identified as GDH coding regions by DNA sequencing. It is interesting to note the difference in size of the fungal product compared with the bacterial and plant products reflecting the presence of an intron within the fungal gene.

Protocol 1. *Continued*

- 2 mM dNTP solution   5 μl
- Primer A (100 pmol/μl)   1 μl
- Primer B (100 pmol/μl)   1 μl
- Genomic DNA   0.1 to 0.5 μg
- *Taq* polymerase[b]   2 units

2. Mix, centrifuge briefly (1 sec) in a microcentrifuge, then overlay with 50 μl mineral oil.

3. Perform PCR reactions using appropriate conditions, usually
   - 95°C, 5 min (initial denaturation),
   - (95°C, 1 min; 55°C, 1 min; 72°C, 1 to 5 min) repeat these cycles of denaturation, annealing and extension 30 to 40 times.
   - 72°C, 2 min to complete DNA synthesis of nascent strands

4. Hold at 4°C until required.

---

[a] 10× PCR buffer: 100 mM Tris–HCl, pH 8.3, 500 mM KCl, 15 mM MgCl$_2$, 0.1% gelatin.

[b] *Taq* polymerase may be added after all the other reactants by pipetting under the mineral oil allowing addition of the enzyme towards the end of the initial (5 min) denaturation step. In practice it seems to make little difference at which point the enzyme is added, and for convenience it is therefore usually added with the other reactants. However, there are some cases where it may be advisable to add *Taq* only the initial denaturation step (see *Chapter 12*, Section 2.1.3)

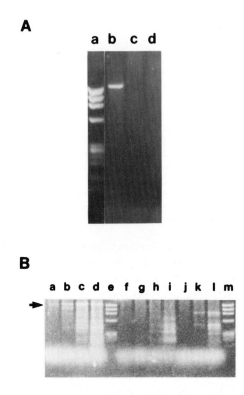

Figure 5. Amplification of a fungal genomic DNA sequence. (**A**) 2% NuSieve GTG agarose gel stained with ethidium bromide. **Lane a** ϕX174 *Hae*III size markers. **Lane b** a 1.4 kbp DNA fragment amplified, according to *Protocol 1*, from genomic DNA of the filamentous fungus *Dactylium dendroides*. Primers 1 and 2 were designed from peptide sequence data derived from galactose oxidase and have potential redundancies of 256- and 512-fold, respectively. **Lanes c** and **d** are single primer controls identical to **lane b** but containing respectively either primer 1 or 2 only. One-tenth of each reaction was loaded.

(**B**) Amplification of *D. dendroides* DNA at a sub-optimal annealing temperature (45°C). **Lane e** and **lane m** ϕX174 *Hae*III size marker fragments. Other tracks show PCR products with different primer combinations. (**a**) to (**d**) primers 1 and 2, (**f**) to (**i**) primer 1 only, (**j**) to (**l**) primer 2 only. In this experiment 5 μl aliquots (one-tenth) of the reaction were withdrawn after 20 (**lanes a** and **f**), 25 (**lanes b, g**, and **j**) and 30 (**lanes c, h**, and **k**) cycles. The reaction was stopped after 35 cycles (**lanes d, i**, and **l**). The expected product is a 1.4 kb fragment (arrowed) visible when both primers are present. A number of other bands are also visible even when only a single primer is present. Increasing the annealing temperature to 55°C produces the result shown in part (**A**).

4.1 Reaction conditions

Optimal PCR conditions can vary for particular reactants and it may be useful to test various reaction conditions and extension times to achieve maximal amplification of unique products. Excellent discussions of these variables and the optimization of the PCR are provided in Chapters 12 and 14. Briefly, it can be useful to check:

(a) *Annealing temperature*; this should be as high as possible, we usually start at 55°C which tends to work well. If possible increase the temperature to increase specificity.

(b) *Magnesium concentration*; if results are poor, titrate from 1 to 5 mM, ideally in 0.5 mM steps.

(c) *Extension time and number of cycles*; the number of cycles required to amplify a particular DNA sequence depends on factors such as gene copy number and genome complexity. A minimum of thirty cycles is usually required, although typically 40 cycles are performed for genomic amplifications. It can be helpful in early experiments to withdraw small aliquots (5 to 10 μl) from the reaction after set numbers of cycles, for example 20, 25, 30, etc. These samples can then be compared by gel electrophoresis to determine the optimal number of cycles of amplification. *Figure 5b* shows the results of such a preliminary experiment under sub-optimal conditions. Multiple bands occur whether single primers or both primers are present, although it is possible to identify a single band that only occurs in the two primer reactions. Optimizing the conditions by simply increasing the temperature from 45°C to 55°C leads to the production of the unique product shown in *Figure 5a*.

4.2 Analysis of PCR products

The simplest and most convenient method for analysis of PCR samples is by electrophoresis through agarose minigels. The agarose concentration should be chosen to allow resolution of DNA fragments within the size range of expected products (approximately 1 to 2% for products >500 bp and 2 to 4% for products <500 bp).

4.2.1 Prediction of expected product size

With a multiple alignment strategy, particularly when amplifying cDNA (which lack introns) it may be possible to predict the expected size of the correct PCR product. If a number of bands are amplified, this prediction may allow identification of the correct band although it is still recommended that the identity of a product be checked by a Southern blot hybridization experiment or by direct sequencing (Section 4.5).

Prediction of the size of a PCR product is not usually possible from peptide sequence from a single protein, unless a peptide map is available. For example, the size of the 1.4 kb fragment (see *Figure 5a*) amplified from primers based on amino acid sequence data from the N-terminus and a protease V8 fragment of the enzyme galactose oxidase, could not be predicted, since the position of the protease V8 cleavage site within the protein was unknown (13).

4.2.2 Primers bands

On some gels an unexpectedly intense band is visible at a position corresponding to approximately 50 bp. This is usually caused by primer dimerization due to limited complementarity between primers. The dimerized product is then efficiently amplified by *Taq* polymerase. In some other cases unincorporated, monomeric primers are visible as a diffuse band.

4.3 Recovery of DNA from agarose

It is usually best to purify amplified DNA fragments for reamplification, DNA sequencing, cloning, or for labelling as hybridization probes. This is achieved by agarose gel electrophoresis using a highly purified agarose such as NuSieve GTG (FMC Bioproducts) and sufficient material for further manipulation can often be recovered from an analytical scale gel. NuSieve gels are more brittle than standard gels, so to prevent tearing of the slots

(a) pour gels in a cold room with the slot former at least 1 mm above the glass plate, and

(b) once the gel is set, flood the surface with electrophoresis buffer before removing the slot former.

If PCR samples, especially those representing the same gene from different species, are to be separated on the same gel, it is best to leave an empty lane between samples to prevent cross-contamination. Keep UV exposure of DNA in the gel to a minimum; leave the gel on its glass plate support whilst viewing. With a razor or scalpel blade, slice the band from the gel in as small a volume of agarose as possible. To avoid cross-contamination use a fresh blade for each band.

Various methods are available for recovering DNA from agarose. Usually, some balance must be struck between high yield recovery and speed and convenience of the method. If yield (~50%) is less important than speed we recommend the use of Spin-X filters (as described in Chapter 12, Section 2.3.3). Glassmilk or Glassfog (Geneclean and Mermaid kits, Bio 101 Inc.) purification used according to the manufacturer's instructions give a higher yield. Glassmilk is specifically designed for fragments less than 1 kb and so is particularly useful for many PCR products.

4.4 Reamplification of PCR products

Where the quantity of material generated particularly from genomic DNA is insufficient for subsequent manipulations a sample of the primary reaction products can be reamplified. Typically 1/100th of the initial reaction is used as template in a PCR under identical reaction conditions to the primary amplification. It is usually best to reamplify a sample of the PCR product recovered from an agarose gel (Section 4.3), although with nested PCR it is sufficient to

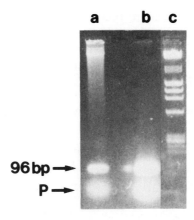

Figure 6. Amplification and reamplification of tomato genomic DNA sequences. A 3% NuSieve GTG preparative agarose minigel stained with ethidium bromide. **Lane c** *Hind*III and φX174 *Hae*III size markers. **Lane a** primary amplification products generated according to *Protocol 1* by 40 cycles of the PCR from 0.5 μg of tomato genomic DNA using primers of 2048- and 8192-fold (see *Figure 1*) potential redundancies. **Lane b** Reamplification products (Section 5.6); one-hundredth of the reaction products from **lane a** were amplified for 30 cycles according to *Protocol 1*. Both the primary and reamplification reactions were ethanol precipitated, redissolved in 8 μl water and loaded on to the gel. The starting genomic DNA is clearly visible in **lane a** and unincorporated primers can be seen at the bottom of the gel (**P**).

use an aliquot of the primary amplification reaction in a second PCR with the nested primer pair. Reamplification products are shown in *Figure 6* lane (b).

4.5 Direct DNA sequencing

Double-stranded PCR DNA can be sequenced directly using the highly degenerate PCR primers as DNA sequencing primers. DNA recovered from agarose by the Spin-X or Mermaid methods (Section 4.3) provide good templates for DNA sequencing reactions using the commercially available kits of Sequenase (US Biochemicals) or T7 polymerase (Pharmacia); detergents can improve sequence data quality (14). It is also possible to sequence DNA directly in melted NuSieve gel (15) and inclusion of DMSO in the sequencing reaction may improve the quality of the data (ref. 16; see also Chapter 4).

Heat denaturation followed by rapid quenching on ice allows sequence to be determined from one stand of the duplex DNA. If the sequencing reactions are performed with dilute labelling mixes and the inclusion of manganese buffer then it is possible to determine the DNA sequence immediately adjacent to the primer. This allows comparison of the DNA coding sequence with the known peptide sequence thus confirming the identity of amplified fragments. In cases where more than one PCR product is generated such an approach may be used to identify the correct fragment prior to further experimentation. In practice, we find that PCR conditions can usually be optimized such that only one product is generated and sequencing then allows confirmation of identity rather than being used to select the correct product from a number of bands. *Protocol 2* describes a simple method for direct sequencing of double-stranded PCR DNA, producing results such as those shown in *Figure 7*.

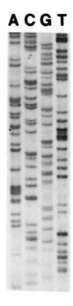

Figure 7. Direct DNA sequencing of PCR DNA. Part of an autoradiogram showing DNA sequence from PCR-DNA product recovered from an NuSieve-GTG agarose gel (*Figure 5A, lane b*) using a Spin-X filter and sequenced according to *Protocol 2*.

Protocol 2. DNA sequencing of double-stranded PCR DNA using Sequenase

1. Mix 0.2 to 2 µg (usually one gel band) PCR DNA recovered from agarose with 2 pmol primer in 8 µl water.

2. Heat to 100 °C for 5 min then rapidly transfer to dry-ice until frozen. Alternatively, chill on wet ice for 5 min.

3. Thaw the solution at room temperature, add 2 µl of 10 × sequencing reaction buffer and continue to incubate at room temperature for 20 min to allow annealing.

4. Dilute the labelling mix 1:20 in double distilled water (1 µl labelling mix + 19 µl water). This solution can be stored at −20 °C for at least one week.

5. Perform the Sequenase reactions according to the manufacturer's instructions, but substituting the 1:20 diluted labelling mix for the normal 1:5 mix, and include 1 µl of manganese buffer per reaction.

6. Fractionate the reaction products on a 6 or 8% acrylamide gel (a wedge gel can be used) run in 1 × TBE^a until the Bromophenol Blue dye is 7 to 8 cm from the bottom.

7. Fix and process the gel as normal. It may be necessary to leave the gel exposed to X-ray film for longer than normal (for 2 to 3 days) to ensure data can be read close to the primer site.

^{a} 10 × TBE stock solution is 108 g Tris base, 55 g boric acid, and 40 ml 0.5 M EDTA (pH 8.0) per litre.

Heat denaturation followed by rapid quenching on ice allows sequence to be determined from one strand of the duplex PCR products. Alternative approaches to sequence determination for PCR products include asymmetric amplification (17), exonuclease digestion of one strand (18) and specific priming on one strand due to the incorporation of a priming site as part of one of the original PCR primers (19).

5. Summary

Highly degenerate primers for use in PCR experiments provide a rapid and direct route to the amplification of regions of essentially any gene for which limited peptide sequence are available. By using multiple sequence alignment data it is also possible to design universal primers that allow the amplification of corresponding regions of gene families from a variety of species. Confirmation of the identity of the PCR product can be derived by direct sequence analysis prior to the use of the PCR product as a homologous hybridization probe to facilitate gene isolation from an appropriate library.

Acknowledgements

We thank Kapil Yadav, Margaret Smallwood, Katherine Lilley, and Jeffrey Keen for amino acid sequence data used to design PCR primers, and Zümrüt Ogel for helpful comments on the manuscript. Financial support for this work was provided by AFRC and a 'grant-in-aid' from Enimont. S.J.G. is supported by a Royal Society University Research Fellowship.

References

1. Lee, C. C., Wu, X., Gibbs, R. A., Cook, R. G., Muzny, D. M., and Caskey, C. T. (1988). *Science*, **239**, 1288.

2. Girgis, S. I., Alevizaki, M., Denny, P., Ferrier, G. J. M., and Legon, S. (1988). *Nucleic Acids Research*, **16**, 10371.
3. Knoth, K., Roberds, S., Poteet, C., and Tamkun, M. (1988). *Nucleic Acids Research*, **16**, 10932.
4. Frohman, M. A., Dush, M. K., and Martin, G. R. (1988). *Proceedings of the National Academy of Sciences, USA*, **85**, 8998.
5. Marmur, J. (1961). *Journal of Molecular Biology*, **3**, 208.
6. McPherson, M. J., Baron, A. J., Jones, K. M., Price, G. J., and Wootton, J. C. (1988). *Protein Engineering*, **2**, 147.
7. Azevedo, M. D., Felipe, M. S., Astolfi-Filho, S., and Radford, A. R. (1990). *Journal of General Microbiology*, **136**, 2569.
8. Jofuku, K. D. and Goldberg, R. B. (1989). In *Plant molecular biology: A practical approach* (ed. C. S. Shaw), pp. 37–66. IRL Press at Oxford University Press, Oxford.
9. Sambrook, J., Fritsch, E. F., and Maniatis, T. (1989). *Molecular cloning: A laboratory manual*. Cold Spring Harbor Laboratory Press, Cold Spring Harbor, NY.
10. Findlay, J. B. C. and Geisow, M. J. (ed.) (1989). *Protein sequencing: a practical approach*. IRL Press at Oxford University Press, Oxford.
11. Schmidt, R. R. (1989). In *New directions in biological control*, UCLA Symposia Molecular and Cell Biology, Vol. 112 (ed. R. Baker and P. Dunn). Alan R. Liss, New York.
12. Sommer, R. and Tautz, D. (1989). *Nucleic Acids Research*, **16**, 6749.
13. Ito, R. N., Phillips, S. E. V., Stevens, C., Ogel, Z. B., McPherson, M. J., Keen, J. N., Yadav, K. D., and Knowles, P. F. (1991). *Nature*, **350**, 87–90.
14. Bachmann, B., Lüke, W., and Hunsmann, G. (1990). *Nucleic Acids Research*, **18**, 1309.
15. Kretz, K. A., Carson, G. S., and O'Brien, J. S. (1989). *Nucleic Acids Research*, **17**, 5864.
16. Winship, P. R. (1989). *Nucleic Acids Research*, **17**, 1266.
17. Gyllensten, U. B. and Erlich, H. A. (1988). *Proceedings of the National Academy of Sciences, USA*, **85**, 7652.
18. Higuchi, R. G. and Ochman, H. (1989). *Nucleic Acids Research*, **17**, 5865.
19. Stoflet, E. S., Koeberl, D. D., Sarkar, G., and Sommers, S. S. (1988). *Science*, **239**, 491.
20. Gurr, S. J., Hawkins, A. R., Drainas, C., and Kinghorn, J. R., (1986). *Current Genetics*, **10**, 761.

12

General applications of PCR to gene cloning and manipulation

TIM CLACKSON, DETLEF GÜSSOW, and PETER T. JONES

1. Introduction

One of the most common applications of PCR is the extraction and manipulation of genes or gene families of known sequence. Such genes can be accessed directly from crude DNA or RNA sources, and amplified by one of a variety of PCR techniques to obtain a suitable product. Subsequent molecular engineering and clone screening steps can also largely be replaced with PCR-based approaches that afford major savings in time, effort, and cost. In particular, DNA segments can be spliced together precisely without the need for convenient restriction sites opening up new possibilities. Of the various 'PCR engineering' techniques that have recently been reported, we have chosen to concentrate on those that are in frequent and successful use in our laboratory. Examples are taken from our experience of cloning immunoglobulin V-region gene segments, and in manipulating whole antibody genes.

2. Cloning genes and gene families from genomic DNA and mRNA

2.1 Genes of known sequence

Armed with PCR technology a researcher can rapidly obtain a clone of any published sequence. The basic principle we have applied when extracting such genes is to ensure the amplification is as stringent as possible, the intention being to generate a single band in quantities that are easy to clone. Since in most cases the entire gene sequence is known long primers can be used, enhancing specificity by allowing the annealing step to be carried out at a high temperature (as an example, see ref. 1). This invariably gives a 'clean' amplification (*Figure 1*) although apparent homogeneity may still mask co-amplification of related genes or pseudogenes. Another advantage of high-temperature annealing is that the fixation of errors is reduced, since mismatches are very poorly extended by *Taq* polymerase at high temperatures (1).

2.1.1 Primer design

It is much more efficient to clone PCR products using unique restriction sites rather than to 'blunt-end' clone. Hence, the primers must incorporate such sites as well as contain target site complementarity. Knowledge of the complete target sequence allows considerable flexibility in the primer design, and some general guide-lines include

(a) The 3' regions of the primers should comprise 25–30 nucleotides of sequence complementary to the target site, to allow high-temperature annealing. If the target sequence must be amplified selectively from a DNA source containing other homologous genes, the 3'-ends should prime in a region of diversity between the regions of homology (see *Figure 1b*).

(b) Unless the restriction site can be incorporated into this 3' primer sequence with only one or two mismatches, it should be appended as a 5' 'tag' to ensure efficient priming during PCR. Any mismatches to the target sequence should ideally be at least 10–12 bases from the 3'-end of the primers.

(c) Some restriction enzymes are very inefficient at cutting sites located at the ends of PCR products. To reduce this problem it is important to incorporate at least six extra nucleotides 5' to the enzyme recognition site.

(d) Other considerations that generally apply to primer design are also valid. In particular, complementary regions which might lead to formation of a 'primer-dimer' (a PCR product derived from a primer-primer annealing event) should be avoided (*Figure 1c* provides an example of this problem). Ideally, the T_m of the primers should be similar, and their (G+C) content should be around 50%.

2.1.2 Preparation of template DNA

If purified genomic DNA or cDNA is available, about 0.1 to 1 μg should be used per amplification for a single-copy gene. However, in nearly all cases genes can be amplified from extremely crude DNA sources. *Protocols 1* and *2* list simple and rapid procedures for obtaining suitable genomic DNA or cDNA essentially by boiling cells. These protocols have been successfully applied to both prokaryotic and eukaryotic cells (*Figures 1* and *2*).

Protocol 1. Crude Preparation of Genomic DNA for PCR

1. Suspend about 10^5 cells (wash eukaryotic cells once in PBS[a]) into 100 μl water and place in a boiling water bath for 5 min.
2. Chill on ice for 1 min, then spin for 3 min in a microcentrifuge at ~13 000 g.
3. Remove the supernatant to a fresh tube. Use 2.5–10 μl in one PCR.

[a] PBS is 25 mM sodium phosphate pH 7.0–7.2, 125 mM NaCl.

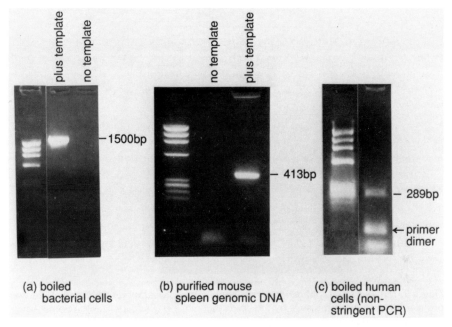

Figure 1. Amplifications of genes of known sequence from genomic DNA. **(a)** and **(c)** show the products of PCR from *E. coli* DNA and human peripheral blood lymphocyte DNA respectively, both prepared by the crude cell boiling technique of *Protocol 1*. Primers were long (~30-mers) to allow annealing at 72°C. In **(c)** annealing was not stringent (45°C) and there is a significant primer dimer. **(b)** shows amplification from purified mouse DNA (*Protocols 4* and *5*) with long oligonucleotides to generate a 'sticky feet' primer (Section 3.3). The target sequence was an immunoglobulin constant region exon with many homologues in the mouse genome. Hence the positions of the 3'-ends of the primers were selected to amplify only the desired gene (Section 2.1.1): the 3' four bases annealed in regions of variation between the homologues.

Protocol 2. Crude preparation of cDNA from eukaryotic cells for PCR

1. Wash approximately 2×10^6 cells (5 ml cultured hybridoma cells) once in PBS.

2. Resuspend cells in 100 μl 0.1% diethylprocarbonate (DEPC) in water (made up immediately before use).

3. Immediately boil for 5 min, then hold on ice for 1 min. Spin for 2 min to pellet cellular debris.

4. Meanwhile, mix the reverse transcription reaction:

   - 5 mM dNTPs pH 7.0   5 μl
   - 10 × first-strand buffer[a]   10 μl

189

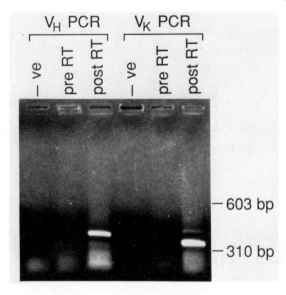

Figure 2. VH and VK amplifications from crude cDNA preparations. RNA and cDNA was prepared from hybridoma cells by the crude technique of *Protocol 2*, with reverse transcription and PCR using the VH and VK primers described in Section 2.2. The 'negative' (no cell boilate added) and 'pre-RT' (only cell boilate added) controls show that the amplification does not originate from contaminating DNA or chromosomal DNA.

Protocol 2. *Continued*

- 0.1 M DTT   5 μl
- Primer (10 pmol/μl)[b]   2 μl

Decontaminate the mix by UV irradiation as described in *Protocol 3*.

5. Add 70 μl supernatant from the boiled cells. Incubate at 65 °C for 5 min, and leave on the bench to cool for 15 min.

6. Add 4 μl RNasin (40 units/μl; Promega Biotech) and 4 μl AMV reverse transcriptase ('super RT', 23 units/μl: Anglian Biotechnology). Incubate at 42 °C for 1 h. Use 5 μl per PCR.

[a] 10 × first strand buffer: 1.4M KCl, 0.5M Tris–HCl pH 8.1 at 42 °C, 80 mM MgCl$_2$ (the correct pH is important).
[b] We usually use the 'forward' PCR primer to prime reverse transcription.

Clearly the choice of RNA or DNA as starting material depends on the experiment. Genes can also been amplified from other complex sources such as DNA purified from a pooled lambda cDNA library.

2.1.3 Amplification

Protocol 3 gives a general method for amplification from genomic DNA or cDNA. The temperature cycling profile should be decided with the following considerations:

(a) *Number of cycles*. This should be determined empirically, starting at 25 cycles. When amplifying from a pooled library or DNA from a low number of cells, 30 or 35 cycles may be required to generate sufficient product for straightforward cloning. In general, the number of cycles should be kept as low as possible to minimize the incorporation of errors (see Chapter 13).

(b) *Extension time*. A useful rule of thumb is to allow one min per kb of DNA; this applies empirically up to about 4 kb. Above this size success also appears to require a high enzyme concentration as well as long extension times.

(c) *Annealing temperature*. With primers designed according to Section 2.1.1, annealing at 72°C should be possible; but in certain cases and especially with primers bearing multiple central mismatches, this may need to be dropped to 65°C or 60°C. Adding the enzyme only at 94°C (as in *Protocol 3*) avoids the non-specific extension that can occur as the reaction mix is heated through non-stringent temperatures: in particular, yields of 'primer-dimer' can be reduced (*Figure 1c*).

(d) *Post-treatment*. To ensure complete extension of all amplified molecules, the cycling should be followed by a 5–10 min incubation at 72°C.

We allow 30–60 sec each for denaturation and annealing (though it is only necessary for the actual tube temperature to reach the set temperature for a few sec). As an example, a suitable cycle to amplify a 1.5 kb sequence using primers which have 25 nucleotides complementarity to the template at their 3′-ends would be:

(94°C for 1 min, 72°C for 2.5 min) × 25 cycles, then 10 min at 72°C.

Protocol 3. General protocol for PCR from genomic DNA or cDNA

1. Mix the following in order:
- H_2O to 50 µl (final vol.)
- 10 × PCR buffer[a]   5 µl
- 5 mM dNTPs (pH 7.0)   2 µl
- Primer 1 (10 pmol/µl)   2.5 µl
- Primer 2 (10 pmol/µl)   2.5 µl

2. Lay the tube on a short-wave transilluminator (254 nm) for 5 min (to inactivate any contaminating DNA; see Section 2.2.5; (8)).

3. Add template DNA, overlay with paraffin (50–100 µl) and place on PCR block at 94°C for 5 min.

Protocol 3. *Continued*

4. Add 0.5 μl *Taq* polymerase (5 units/μl) under the oil and start temperature cycling.

5. Analyse 5–10 μl on an agarose gel.

[a] 10 × PCR buffer: 0.1 M Tris–HCl pH 8.3 at 25°C, 0.5 M KCl, 15 mM MgCl$_2$, 1 mg/ml gelatine (for the Cetus enzyme).

2.1.4 Trouble-shooting

In all cases using the long primers described above we have found that a failure to amplify can be solved simply by adjusting the annealing temperature: if 72°C results in failure, try 65°C and then 60°C. The main other variable is the Mg^{2+} concentration; however, at 1.5 mM MgCl$_2$ (as in the standard PCR buffer) nearly all amplifications are efficient. A MgCl$_2$ titration between 1 and 10 mM final concentration can then be performed to optimise conditions if a highly efficient PCR is required.

If primer-dimer problem persists even when adding the enzyme at 94°C, it may be worth adding one primer and then the other, all at 94°C. In the very rare cases when a given primer pair will not work, redesigning them by shifting the priming site by a few bases is likely to solve the problem.

2.2 Gene families

2.2.1 Introduction

In antibodies, a heavy and light chain variable domain (VH and VL respectively) constitute the antigen binding region. VH and VL genes are constructed by recombination at the genetic level between a large number of gene fragments, so that each B cell expresses on unique VH and one unique VK gene (most mouse light chains are kappa, K). We have been using PCR to clone libraries of mouse VH and VK genes (for background see refs 4 and 5 and references therein). The genes are about 400 bp long and primers have been designed to anneal near the ends, in regions of sequence of conservation. PCR has been performed on DNA and RNA isolated from spleen cells and peripheral blood lymphocytes (PBLs) of immunized and naïve animals.

VH and VK genes are an atypical gene family in that an individual B cell encodes only a single family member. However, the primer design and template preparation protocols described here are equally applicable to other cases where coamplification of related genes is required. For a review of several other gene families that have been accessed by PCR, see ref. 3.

2.2.2 Primer design

Although we have found the following guide-lines useful when designing primers to amplify gene families, perhaps the most important step is simply to try several different primer systems. A degree of optimization is likely to be necessary in order to amplify a large proportion of the gene family.

(a) *Identifying conserved regions.* Mouse VH and VK genes are also unique in that a large sequence database (that of Kabat, (6) containing several hundred sequences) is available. In this case a computer program (FAMNS: R. Staden, unpublished data) was used to locate conserved regions. For other families it may be sufficient to manually align known sequences and inspect by eye to derive a consensus sequence. At some positions in a conserved region two nucleotides may be equally common and this can be reflected in a partially degenerate primer sequence. *Figure 3* illustrates the steps involved in the design of the VH primers VH1FOR and VH1BACK. A similar approach was taken with VK primers VK1FOR and VK1BACK. All four primers are unique except for VH1BACK which is 32-fold degenerate (4).

(b) *Restriction sites.* As with amplification of unique genes, restriction sites can be incorporated either within the region of primer–template complementarity, or as a tag near the 5'-end. The choice depends on the nature and extent of the conserved region chosen as a priming site. With gene families, maximizing complementarity of the primers to the template genes is more important than engineering a restriction site. Suitable regions for introducing restriction sites can be identified as follows:

(i) Confine the selection of restriction enzymes to those that cut rarely in the known family sequences. It is also advisable to concentrate on rare-cutter enzymes such as 'eight cutters' or those with recognition sites in common vector polylinkers.

(ii) Identify any near matches within two nucleotides to one of these sites in the conserved region. The one or two mismatches to the consensus sequence that are required to engineer a site here will probably not reduce the specificity of the primer, as long as substantial (at least 10 nucleotides) stretches of good complementarity flank the region.

(iii) Alternatively, a restriction site can be incorporated over a short stretch of variable sequence within the conserved region, since primer–template complementarity will already be lower in this area.

(iv) If recognition sites cannot be incorporated centrally, they should be appended at the 5'-end of the primers. As described in Section 2.1.1, at least six bases should be included 5' to the restriction site to facilitate efficient cleavage by the enzyme.

For the VH and VK primers the approaches outlined in (ii) and (iii) were applied, with sites being engineered over regions of poor complementarity but which already contained some of the correct sequence for the site (*Figure 3*).

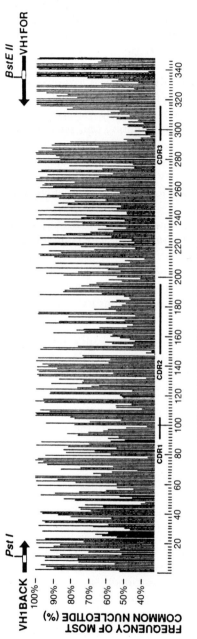

Figure 3. Design of primers to amplify the mouse VH gene family. The known sequences of mouse VH genes were used to plot the frequency of the most common nucleotide, to identify consensus sequences for primer annealing. The lower panel shows how restriction sites were incorporated to minimize mismatches with the sequence consensus. Mouse VH and JH segments (from which the two primer binding sites are derived) can be subgrouped into families, and their consensus sequences alignments are compared to the forward and backward primers. Nucleotides not matching the primer sequences are highlighted in bold.

(c) *Primer length.* The most important feature of the primers is high complementarity at the 3'-end. The 3'-ends should therefore be designed to anneal in the region of highest sequence conservation over the greatest possible length (at least 15 bases). The overall length will depend on the extent of sequence conservation; the VH and VK primers were between 22 and 34 bases.

(d) *Universal or family specific primers?* Such sequence comparisons can be used to design unique universal primers, which would be expected to amplify nearly all gene family members. In the case of VK and VH genes, sequences can be subgrouped into families and a primer designed for each subfamily (6) (*Figure 3*). A pool of such primers can therefore be used in PCR and may give access to sequences sufficiently divergent to be missed by the universal primers (4).

It is informative to compare the final primer sequences against each known individual gene sequence in the database, particularly to check for sequences that have more than two mismatches in the first fifteen 3' nucleotides of the primers. Such a scan can reveal subfamilies of sequences that may be under-represented in the PCR product and suggest modifications to the primer design (4).

2.2.3 Template preparation

(a) *High-quality DNA and cDNA preparations.* We have successfully amplified DNA and total RNA isolated from spleen cells and PBLs. Efficient fractionation procedures for these cells are described in *Protocol 4*. From these cells, we often obtain acceptable results using the crude DNA and cDNA templates prepared as in *Protocols 1* and *2*.

Protocol 4. Preparation of lymphocytes for PCR

A. *Preparation of human peripheral blood lymphocytes (PBLs) from blood.*

All steps are carried out at room temperature. Care should be exercised when handling human material.

1. Dilute 20 ml heparinized blood with an equal volume of PBS.

2. Carefully layer 20 ml on to 15 ml 'cushions' of Ficoll Paque (Pharmacia) in two Falcon 2098 tubes or similar.

3. Spin for 20 min at 780 g (maximum).

4. Harvest the PBLs from the interface with a Pasteur pipette.

5. Dilute in an equal volume of PBS, and repellet at 400 g (maximum) for 5 min.

6. Aspirate away the supernatant and wash again in 1 ml PBS. Resuspend in 50 μl PBS (expect a yield of about 10^8 cells).

Protocol 4. *Continued*

B. *Preparation of lymphocytes from a mouse spleen.*

1. Cut the spleen into small pieces and transfer to a Potter homogenizer. Homogenize with a loose-fitting piston.

2. Transfer the suspension to a 15 ml Falcon tube avoiding carry-over of connective tissue and dilute to 15 ml with PBS. Wash several times by pelleting at low speed (400 g for 5 min) in a bench centrifuge and re-suspending in PBS.

3. (optional) Enrich for lymphcytes using a Ficoll cushion as described above for PBLs, but spin for 30 min and wash several times to remove Ficoll.

4. Resuspend in 500 μl PBS; expect a yield of about 10^8 cells. At this stage the cells can be stored in aliquots in liquid N_2, cooling slowing (over-night in a sealed box at −70°C), in 80% fetal calf serum/20% DMSO at 10^6 cells/ml.

However, to amplify diverse libraries a highly efficient PCR is required to increase the chances of obtaining genes poorly matched to the primer, and it is worth preparing highly purified DNA or RNA. The preparation of suitable genomic DNA from eukaryotic cells is described in *Protocol 5*, and the preparation of cDNA in *Protocol 6*. PCR using material prepared in this way yields excellent results (*Figure 4*). The main disadvantage of using purified template is that more handling is required and this increases the chances of contamination (see Section 2.2.5).

Protocol 5. Preparation of high-quality genomic DNA from eukaryotic cells for PCR

1. Resuspend the cells (1–5 × 10^7; about half the cells from the fractiona-tions in *Protocol 4*) in 500 μl PBS and wash twice in PBS, by pelleting in a microcentrifuge at low-speed setting (~6000 r.p.m.) for 1 min. Resuspend in 200 μl PBS.

2. Add the suspension to a Falcon 2059 tube (or similar) containing 2.5 ml 10 mM Tris–HCl pH 7.4, 10 mM EDTA with slow vortexing.

3. While still vortexing add 25 μl 20% SDS. The cells should lyse immediate-ly and the solution become very viscous and clear.

4. Add 100 μl proteinase K (Boehringer; 20 mg/ml) and incubate 1–3 h at 50°C.

5. Transfer the lysate to a Falcon 2098 tube and extract with an equal vol. of phenol, then an equal volume of chloroform. Vortex, spin to separate the phases, and transfer the upper aqueous layer to a fresh tube.

6. Overlay with 3 vol. of ethanol and mix gently until the DNA is visible.[a]

Flame the end of a Pasteur pipette to seal it then spool out the DNA and let the ethanol drain off.

7. Redissolve the DNA in 1 ml TE[b] at 4°C overnight on a roller. Expect about 0.5–2 mg DNA.

[a] It is important not to precipitate the DNA in ethanol for too long as this will dehydrate the DNA and it will take days to redissolve.
[b] TE: 10 mM Tris–HCl pH 8.0, 0.1 mM Na_2 EDTA.

Protocol 6 is a modification of reference 20 and gives excellent results with spleen and hybridoma cells: addition of VRC (vanadyl ribonucleoside complex) as an RNase inhibitor is necessary for spleen cells (*Figure 4*). Guanidinium isothiocyanate/CsCl procedures (21) yielding total cellular RNA also give good results but are more time-consuming.

Protocol 6. Preparation of high-quality total cytoplasmic RNA/cDNA from eukaryotic cells for PCR

1. Wash 1 to 5×10^7 freshly harvested cells in 50 ml PBS at 800 g for 10 min.

2. On ice, add 1 ml ice-cold lysis buffer[a] to the pellet and resuspend it with a 1 ml Gilson pipette. Leave on ice for 5 min.

3. Spin for 5 min at 4°C in a microcentrifuge at 13 000 r.p.m., in pre-cooled tubes.

4 Transfer 0.5 ml of the supernatant to each of two Eppendorfs containing 60 µl 10% (w/v) SDS and 250 µl phenol (equilibrated with 100 mM Tris–HCl pH 8.0). Vortex hard for 2 min, then microcentrifuge (13 000 r.p.m.) for 5 min at r.t. Phenol *must* contain 8-hydroxyquinoline.

5. Re-extract the aqueous upper phase 5 times with 0.5 ml of phenol.

6. Precipitate with 1/10 vol. 3 M sodium acetate and 2.5 vol. ethanol at 20°C overnight or dry-ice/isopropanol for 30 min.

7. Wash the RNA pellet and resuspend in 50 µl water. Use 2.5 µl to check concentration by measuring OD_{260} and check 2 µg on a 1% agarose gel (see *Figure 4a*).

8. Set up the following reverse transcription mix:
   - H_2O (DEPC-treated)   20 µl
   - 5 mM dNTP   10 µl
   - 10 × first-strand buffer (*Table 2*)   10 µl
   - 0.1 M DTT   10 µl
   - Primer(s) (10 pmol/µl)   2 µl of each
   - RNasin (Promega; 40 units/µl)   4 µl

9. Dilute 10 µg RNA to 40 µl final volume with DEPC-treated water. Heat at 65°C 3 min and hold on ice 1 min to remove secondary structure).

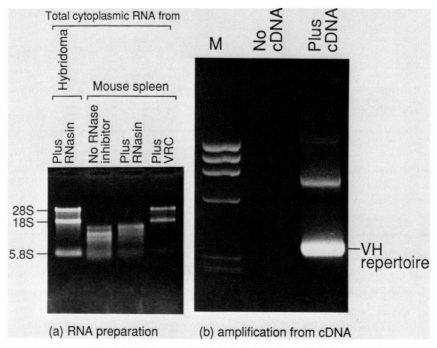

Figure 4. VH amplification from high-quality mouse spleen cDNA. **(a)** shows total RNA (2 µg) preparations as in *Protocol 6* demonstrating the importance of using VRC (vanadyl ribonucleoside complex) during cell lysis: RNasin is not sufficient. Amplification (shown in **b**) from cDNA prepared as in *Protocol 6* generates a diverse VH repertoire.

Protocol 6. *Continued*

10. Add to the RNA the reverse transcription cocktail and 4 µl 'Super RT' (Anglian Biotechnology) and incubate at 42°C for 1 h.

11. Boil the reaction mix for 3 min, cool on ice for 1 min, and then spin in a microcentrifuge to pellet debris. Transfer the supernatant to a new tube and use 10 µl per PCR.

ᵃLysis buffer: 10 mM Tris–HCl pH 7.4, 1 mM MgCl$_2$, 150 mM NaCl, 10 mM VRC (New England Biolabs), 0.5% (w/v) Triton X-100, prepared fresh.

(b) *DNA or RNA?* Clearly the choice depends on the gene system under study and the intron structure. Starting from cDNA avoids the coamplification of (most) pseudogenes, and in the case of V-regions this includes the unproductively-rearranged allele present in many B-cells. If amplifying from cDNA, we usually include a control amplification from RNA without reverse transcription, to assess the contribution of DNA-derived amplification to the final product. This does not usually constitute a problem even with the crude 'boiling' technique of *Protocol 2* (*Figure 2*).

2.2.4 Amplification

Amplification (and trouble-shooting) should be carried out as described in Section 2.1.3. Trials should be carried out to ascertain the optimum annealing temperature: for the VH and VK primers, annealing at 60 °C gives a diverse repertoire of V-region amplification products shown by the size heterogeneity of the band but with little non-specific amplification (*Figure 4*).

2.2.5 Problems with contamination

In a busy cloning laboratory working on a particular gene family, contamination of PCR reactions with previously cloned material quickly becomes a major problem. It is therefore essential to run a negative control (omitting template DNA) with every amplification. Observance to the following precautions will reduce the problem to a minimum (ref. 7; see also Chapter 3).

(a) Keep pre-PCR materials (reagents, pipettes and tips, racks etc) seperate from cloned and amplified DNA. PCR reactions should be set up in a separate room not used for DNA work (we use a tissue culture laminar flow hood).

(b) Confine one set of pipettes exclusively for use in setting up PCRs. If possible use positive displacement pipettes (e.g. Microman, Gilson); these pipettes remove the danger of aerosol-mediated transfer of DNA from the pipette barrel. Depurinate contaminated barrels with 0.25 M HCl.

(c) Aliquot all solutions and primers into small volumes in the 'DNA-free' area using disposable plasticware. Discard all current aliquots when contamination is detected.

(d) Wear gloves and change them frequently.

(e) Irradiate the reactions with short-wave UV prior to adding template DNA (ref. 8; as described in *Protocol 3*).

2.3 Digestion and cloning of PCR products

After checking on an analytical gel (*Protocol 3*), the remaining material should be extracted once with chloroform (to remove the oil), then twice with phenol/chloroform. DNA can be precipitated with 0.5 vol. 7.5 M NH_4Ac and 2.5 vol. ethanol, and spinning immediately for 10 min in a microcentrifuge at ~13 000 g. The pellet is washed in 1 ml ethanol, dried and resuspended in 20 µl TE.

2.3.1 Restriction digestion

It is often advisable to perform digestions of PCR fragments in the universal potassium glutamate buffer, 1 × KGB (9). In our experience some enzymes (such as *XhoI*) cut sites near the ends of PCR fragments far more efficiently in this buffer than those containing NaCl. $10 \times$ KGB buffer is 1 M potassium glutamate, 0.25 M Tris–acetate pH 7.6, 0.1 M MgCl2, 0.5 mg/ml BSA, 5 mM 2-mercaptoethanol. Clearly a buffer supplied by manufacturer will give optimal digestion. In either case, efficient cutting usually requires a large overdigestion (at least 20 units of enzyme for at least 3 hours).

2.3.2 Gel purification

The digests should be purified on an LGT agarose gel using TAE as the running buffer. To disrupt protein-DNA complexes, we include 0.2% SDS in the sample buffer and supplement the running buffer with EtBr to 1 μg/ml. For optimum yields DNA should be isolated from the appropriate gel slice using Geneclean (Bio 101 Inc.) according to the manufacturers' instructions. For fragments of less than a kilobase, centrifugal filtration through a Spin-X column (Costar) is a convenient alternative. Gel material is retained on a cellulose acetate filter cartridge and the DNA emerges in the filtrate:

(i) Load the gel slice into the cartridge of a Spin-X column.

(ii) Freeze the entire tube in dry ice for 10 min and thaw. Repeat (optional). This procedure breaks down the gel matrix.

(iii) Spin at ~13 000 r.p.m. in a microfuge for 15 min, preferably in a horizontal rotor to use the whole surface of the membrane.

(iv) Add 1/10 vol. 3 M sodium acetate and 2.5 vol. ethanol to the filtrate, chill on dry ice for 5 min, and spin at ~13 000 r.p.m. for 10 min. Wash the pellet in 1 ml ethanol, dry under vacuum and take up in 10–20 μl water or TE.

3. Manipulating and mutating cloned genes

In this section several PCR-based techniques are described for introducing directed mutations into DNA, or for joining unrelated sequences together. In the simplest cases, point mutations or even short regions of complete degeneracy can be engineered at the priming sites using mismatched primers, followed by recloning of the PCR product as a restriction fragment (Section 3.1). Alternatively, whole plasmids can be amplified with a change incorporated, then recircularized by ligation (Section 3.2). Both of these approaches offer advantages of speed and efficiency over conventional mutagenesis protocols. Furthermore, stretches of new sequence can be added to the ends of PCR products by using primers with 5' 'tags'. These terminal extensions can be used to join the amplified region to completely unrelated sequences to which the tags are complementary. Such techniques (Sections 3.3 and 3.4) permit precise cutting and pasting of two or more sequences without the need for convenient restriction sites.

3.1 Altering flanking sequences by reamplifying cloned DNA

3.1.1 Incorporating point mutations

Using PCR it is simple to regenerate an insert flanked by new restriction sites or other point mutations (*Figure 5a*). The concentration of competing non-target sequences in cloned DNA (including bacterial colonies and plaques) is low compared to PCR from genomic DNA, so that PCR conditions and primer design are less critical. It is still advisable to place mismatches some distance from the 3'-ends of the primers; however, we have efficiently amplified from an M13 plaque using a 17-mer primer mismatched at three positions.

In general, the following guide-lines are useful when amplifying from cloned DNA:

(a) Start with a large amount of template DNA (a toothpicked plaque or colony; Section 4, or 10 to 100 ng of purified DNA) and amplify for only 10–15 cycles. This minimizes the chances of error incorporation.

(b) Any primers of 17 nucleotides or longer can be used. Useful rules of thumb for deciding the appropriate annealing temperature are:

- 17–19-mers   55°C
- 19–21-mers   60°C
- 21–24-mers   65°C
- >24-mers   72°C

(c) Allow 1 min extension time per kilobase amplified DNA.

These guide-lines still apply to primers bearing limited mismatches (especially away from the 3'-end). *Figure 5a* illustrates an experiment to change the restriction sites flanking the VK sequences in a cloning vector. Annealing was performed at 65°C.

(a) Replacing restriction sites.

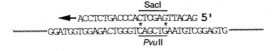

(b) Inserting short stretches of random sequence.

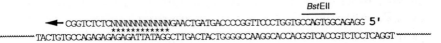

Figure 5. Design of mismatched primers for altering the sequences flanking inserts. In both cases, restriction sites are distanced 6 bases from the 5'-end, and mismatches by at least 9 from the 3'-end. In **(a)** annealing could be performed at 65°C, but in **(b)** 40°C was used due to the large mismatched region near the 3'-end. * indicates a mismatch.

3.1.2 Incorporating short stretches of random mutations

A region of complete degeneracy can be introduced near the end of a PCR product using primers of mixed sequence at one or more positions (10) as illustrated in *Figure 5b*. In this case the primers were of different lengths but annealing at 40°C gave an efficient PCR.

Digestion and cloning of these PCR products can be accomplished as in Section 2.3.

3.2 'Inverse' PCR (IPCR) mutagenesis

In this procedure, a sequence change in a plasmid is directed by one of two apposed primers which anneal 'back-to-back' (*Figure 6*). PCR results in amplification of the entire plasmid with the change now incorporated, and this product can then be purified and self-ligated to regenerate a circular vector. The procedure is extremely rapid and 50–100% of resulting colonies harbour the mutant sequence (ref. 11; see also Chapter 9).

3.2.1 Experimental procedure

The general method is described in *Protocol 7*. Note in particular:

(a) The original protocol (11) can produce deletions at the ligation point. This does not occur if the oligos are kinased before PCR. Clearly, in some cases it may be possible to use a restriction site shared by the primers to effect re-ligation.

(b) Often the original plasmid will co-purify with the IPCR product on a gel. A good yield of mutants therefore depends on amplifying from a low concentration of starting DNA.

(c) For unknown reasons it is necessary to use abnormally long extension times for total plasmid PCR; we normally use 6 mins.

(d) The *Escherichia coli* DNA polymerase I large fragment (Klenow) treatment removes the extra nucleotide (usually A) frequently added at the 3'-end of PCR products, which otherwise prevents blunt-end ligation (11).

Protocol 7. Inverse PCR mutagenesis

1. Set up the following reaction mixes, to kinase the primers:
   - H$_2$O   2.5 µl
   - 10 × kinase buffer^a 2 µl
   - 5 mM ATP   4 µl
   - 0.1 mM DTT   1 µl
   - primer (10 pmol/µl)   10 µl
   - Polynucleotide kinase (10 units/µl)   0.5 µl

   Incubate 37°C for 30 min, then 94°C 5 min to inactivate the kinase.

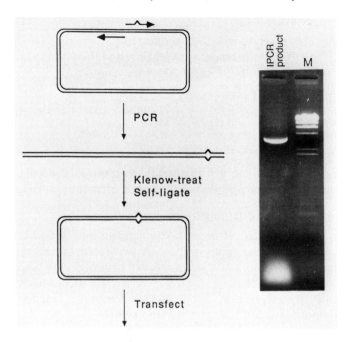

Figure 6. Inverse PCR mutagenesis. The gel shows the products of IPCR from a ~4 kb pUC-based plasmid using long primers (30 bases of 3' complementarity, annealing at 72°C). The spike indicates a point mutation or insertion with respect to the starting plasmid.

2. Set up the following PCR:

- H₂O   to 20 μl

H_2O	to 20 μl
10 × PCR buffer (*Protocol 3*)	2 μl
5 mM dNTPs	1 μl
Primer 1 (kinased, 5 pmol/μl)	1 μl
Primer 2 (kinased, 5 pmol/μl)	1 μl
Plasmid DNA	1–10 μl[b]
Taq polymerase (5 units/μl)	0.5

3. Perform 25 cycles of 94°C 1 min, 45–55°C 1 min, 72°C 6 min. Incubate at 72°C for 10 min after cycling.

4. Purify the products through 1% low-melting-point agarose (TAE buffer), and elute the fragment into 5 μl water (use Geneclean, according to manufacturer's instructions).

5. Make up to 10 μl in 1 × TM[c] buffer with 250 μM dNTPs and 2.5 units Klenow. Incubate for 30 min at 37°C, then 75°C for 15 min.

Protocol 7. *Continued*

6. Use 2 μl of this reaction in a 10 μl ligation mix containing 1 mM ATP and 200 units T4 DNA ligase. Incubate 4 h at r.t. and then use 5 μl to transfect suitable component cells[d]. Expect 200–1000 colonies.

[a] 10 × kinase buffer: 0.5 M Tris–HCl pH 8.0, 0.1 M MgCl$_2$.
[b] Use 0.1–1 fmol plasmid DNA as template; for example, a 1:10000 dilution of a miniprep.
[c] 10 × TM buffer: 0.1 M Tris–HCl pH 8.0, 0.1 M MgCl$_2$.
[d] Use a restriction-defective strain; the DNA will be unmethylated.

Primers are typically 30-mers, with mismatches positioned at least 10 nucleotides from the 3′ end. Such primers usually anneal efficiently at 55 °C or higher.

3.2.2 Applications and limitations

IPCR can be used to engineer limited base substitutions and insertions, and also deletion mutants (the priming sites are separated by the region to be deleted). Larger insertions can be accomplished by using primers with large 5′ 'tags'. However, the technique is clearly restricted to plasmids of a size tractable to PCR; in practice, about 4–5 kb. We have found it ideal for constructing mutations in small pUC-based vectors.

3.3 'Sticky feet'-directed mutagenesis

'Sticky feet'-directed mutagenesis is a method for precisely inserting large sections of DNA into an unrelated template sequence, without using restriction sites (12). The method is based on the principle of site-directed mutagenesis (*Figure 7*). A PCR product, containing the sequence to be inserted, is used as a long primer to anneal to a single-stranded template. This primer anneals precisely due to 'sticky feet' sequences (complementary to the template), introduced at the ends of the long primer from the PCR oligonucleotides. It is not necessary to isolate the single-strand of the PCR product which anneals to the template as an efficient annealing regime has been developed for the double-stranded PCR species. Hence the entire procedure is simple and rapid. After transfecting the extension mix, 25–45% of M13 plaques are usually mutant, and can be identified by PCR screening (Section 4) or sequencing.

The technique has been used to exchange exons of immunoglobulin genes (about 400 bp), and also sections of DNA of completely different sequences and sizes. Hence, it is a general method for large insertion or replacement mutagenesis, in which the junction points between the two sequences are dictated entirely by the design of the PCR oligonucleotides. The procedure is conveniently divided into four sections; PCR to generate the primer, template preparation, mutagenesis, and screening.

3.3.1 PCR conditions and oligonucleotide design

The 3′-ends of the oligos are designed to amplify the region to be transplanted

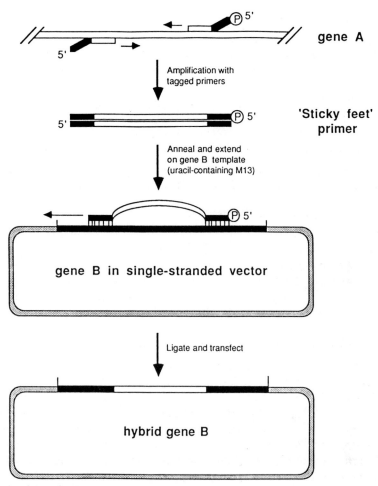

gene A

'Sticky feet' primer

gene B in single-stranded vector

hybrid gene B

Figure 7. 'Sticky feet'-directed mutagenesis.

into the template while the 5'-ends contain 'tags' of 25–30 nucleotides complementary to the template. These tags correspond to the 'sticky feet' in the final PCR product. The length of the 3'-ends depends on the type of DNA to be used as PCR substrate. For deriving a 'sticky feet' primer directly from genomic DNA or cDNA, the principles discussed in Section 2.1 apply; use 25–30 nucleotides and anneal at 72°C during PCR. Though this makes the whole oligonucleotide long (~50 nucleotides), highly specific amplifications are obtained (*Figure 1b*). If amplifying from cloned DNA sources such as plasmid DNA or a phage plaque, it is sufficient to have 15–18 bases of 3' complementarity and anneal at 45°C, since the long 5' tag decreases the effective T_m of the oligo. Otherwise, the standard principles governing PCR apply (Section 3.1); use the minimum number of cycles possible (10–15 from cloned DNA) to reduce the chances of errors.

In practice, the easiest way to design oligonucleotides for 'sticky feet'-directed mutagenesis is to write down the sequence of the desired product. The required oligonucleotide sequences can then be read directly from the junction regions between the two sequences (12).

3.3.2 Preparation of template

By growing the template in a *dut ung* strain, uracil is incorporated in place of thymine at a low frequency. This template strand is selectively destroyed after transfection of a heteroduplex into a wild type (ung^+) strain, providing a powerful selection against this strand and therefore for mutants (13). A suitable strain for growing M13 (except those harbouring an amber mutation) and phagemid templates is CJ236 (13). Simple preparations of M13 (13) and phagemid (14) uracil-containing templates have been described.

3.3.3 Mutagenesis

Protocol 8 details the procedure which applies to both M13 and phagemid templates. The annealing procedure involves a *Taq* polymerase extension during cooling, which gives efficient annealing despite the presence of the competing complementary strand to the long primer. However, extension is completed at a low temperature by another polymerase, that of phage T7.

Protocol 8. 'Sticky feet'-directed mutagenesis

Set up a control with no primer to check for spurious priming from template contaminants, and a control using an oligonucleotide primer (such as the M13 forward sequencing primer) to check the efficiency of extension/ligation.

1. Kinase the forward PCR primer as in *Protocol 7* (final concentration 5 pmol/μl).

2. Perform PCR using the kinased forward oligo and the backward oligo, on appropriate DNA substrate (see Section 3.3.1). It may be convenient to set up several reactions to provide enough long primer for several experiments.

3. Gel-purify the PCR product through low-melting-point agarose (use TAE buffer), and excise the band.

4. Elute the band from a gel slice; this is conveniently performed by Spin-X filtration (Section 2.3.2). Redissolve the DNA in 13 μl H_2O per original PCR reaction. Klenow treatment may be advisable (see *Protocol 7*).

5. Set up the following anneal mix in a 0.5 ml tube:
   - PCR product   6.5 μl
   - M13/phagemid single-stranded template grown in a *dut ung* strain (0.2 μg/μl)   1 μl

- 5 mM dNTPs   1 µl
- 10 × PCR buffer (*Table 3*)   1 µl
- *Taq* polymerase (5 units/µl)   0.5 µl

Overlay with paraffin and place on a block pre-heated to 92°C.

6. Incubate at 92°C for 3 min, 67°C for 0.5 min, and 37°C for 1 min, with about 1 min to drop between each step (ramp 2 on a Techne PHC-2 machine).

7. During the annealing, make up the following extension-ligation mix:
   - H$_2$O   5.7 µl
   - 10 × T7 buffer[a]   1 µl
   - 5 mM ATP   1 µl
   - 0.1 M DTT   1 µl
   - T4 DNA ligase (400 units/µl)   1 µl
   - T7 DNA polymerase (4 units/µl)   0.3 µl

8. Add this mix to the annealing mix immediately after the 37°C step. Incubate 20 min at r.t., then add 1 µl 0.5 M EDTA and 30 µl H$_2$O.

9. Use 0.1, 1 and 10 µl of this reaction to transform a *dut*$^+$ *ung*$^+$ strain of *E. coli* such as TG1.

10. Screen for mutants as described in Section 3.3.4.

[a] 10× T7 buffer: 0.4 M Tris–HCl pH 7.4, 0.2 M MgCl$_2$, 0.32 M NaCl, 50 mM DTT.

3.3.4 Screening for mutants

Yields of mutants are often sufficient to allow screening directly by sequencing. However, we recommend identification of mutant plaques or colonies first by PCR screening (Section 4), especially if the sequences swapped are of totally different lengths, when the yield may be lower. In such cases, a PCR screen with flanking primers will yield a product of diagnostic size. Alternatively, use a single primer and one specific for the inserted sequence, so that only desired mutants give a product. One or two positive clones grown up during the PCR screen (see Section 4.4) can then be sequenced as a check. It is advisable to plaque purify M13 clones to avoid mixed sequences.

3.4 Gene assembly by 'splicing by overlap extension'

If two PCR products bear overlapping complementary ends, they may prime on each other and be extended to yield a hybrid product. A second PCR with two primers annealing at the non-overlapping ends will amplify this hybrid, which can then be cloned if required (15). This general principle of 'splicing by overlap extension' ('SOE') can be applied both to rejoining two halves of the same gene with a sequence change incorporated into the overlap (i.e. site-directed mutagenesis), or to joining two completely unrelated sequences (ref. 16; *Figure 8*). Indeed, it is possible to assemble three separate overlapping fragments in a single assembly PCR (*Figure 8*). In all cases it is necessary to purify the 'primary' PCR products away from the primers before the second step.

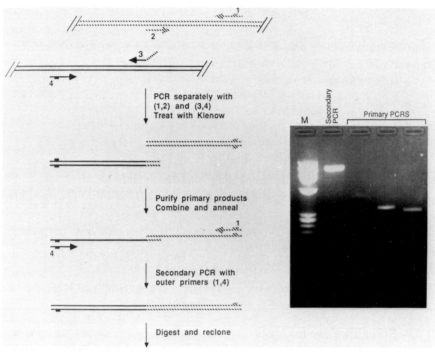

Figure 8. 'Splicing by overlap extension'. The diagram illustrates the joining of two unrelated sequences. The gel shows the primary and secondary PCR products in an experiment to assemble three overlapping fragments. The overlaps were 24 bases and the secondary PCR (all three primary products mixed together) involved 12 cycles, annealing at 72°C. The filled blocks indicate restriction sites incorporated from the primers to allow cloning.

3.4.1 Primer design

As with 'sticky feet'-directed mutagenesis, the primary PCRs will involve primers with 5' tags. We have found it best to generate overlaps of 24 nucleotides in the 'primary' PCRs. This means that the second PCR can be highly stringent (annealing at 72°C), giving a clean and efficient assembly (*Figure 8*) and lowering error rates (Section 2.1). Again, the length of the 3' regions to amplify the target DNA depends on the nature of the template DNA; 15–18 nucleotides for cloned DNA (Section 3.1) and 25–30 nucleotides for genomic or cDNA (Section 2.1). *Protocol 9* describes the general procedure.

3.4.2 Primary PCRs

It is essential to treat the PCR products with Klenow to remove extra 3′ bases (Section 3.2.1) that can otherwise be incorporated at a high frequency into the reassembled product. Proofreading polymerases such as 'Vent' (New England Biolabs) may not suffer from this problem.

3.4.3 Purifying the primary PCR products

We gel-purify PCR fragments destined for assembly (removing both excess primers and template DNA), and elute from agarose using Spin-X columns (Costar; see Section 2.3.2) for fragments smaller than 1 kb, or Geneclean (Bio 101 Inc.) for larger fragments. Alternative methods, such as centrifugation through a concentration filter (15) or ammonium acetate precipitation, do not remove template DNA and are not suitable if spurious bands appear in the primary PCR, a possibility after amplifying from genomic DNA.

3.4.4 Error considerations and limitations

It is very important to minimize the overall number of cycles used, or the accumulation of errors in the assembled product will be unworkably high (see Chapter 13). Primary PCRs from cloned DNA, where the template DNA concentration is high, will yield sufficient product after only ~15 cycles. Similarly, using a significant proportion of these purified bands in the second PCR means that only about 10 cycles need be performed. Note that performing PCR annealing at 72°C also reduces the fixation of errors, since mismatch extension is very poor at high stringency (1). Nevertheless, in practice it is essential to sequence right through the amplified region.

Although errors are still a problem, SOE has become a more attractive technique as PCR conditions have been optimized. It is most useful for precisely joining several unrelated sequences, and this process is high efficient in a single assembly step (rather than stepwise as in ref. 16) if this assembly is carried out at high stringency.

Protocol 9. Splicing by overlap extension

1. Perform primary PCR reactions under conditions described in Section 3.4 to minimize errors. After amplification, add 1/10 vol. 5 mM dNTPs and 2.5 μl Klenow (5 units/μl) and incubate at 37°C for 10 min.

2. Gel-purify the products and elute using Geneclean or Spin-X into 10 μl water (Section 2.3.2).

3. Perform secondary PCR using 2.5 μl of each primary product and annealing at high temperature if possible, for 10 cycles.

4. Digest and clone the assembled product (Section 2.3).

---

4. Characterizing recombinant clones directly by 'PCR screening'

Bacterial colonies and plaques, transferred directly by toothpick, are efficient PCR substrates. This allows the recombinant phage or plasmid vectors they harbour to be analysed directly by PCR, without the need for any DNA preparation. By an appropriate choice of primer pair, the presence, size and orientation of the inserts in any number of clones can be rapidly established (17). 'PCR screening' has largely replaced minipreps and restriction analysis for clone characterization in our laboratory.

The general procedures for screening are described in *Protocol 10*, and examples are shown in *Figure 9*. Once material from the plaque or colony has been transferred into the PCR mix, viable cells remain on the toothpick. These may be used to 'rescue' the clone by innoculating a small-volume culture for subsequent use, for example to prepare sequencing template. The culture may be grown during PCR, so that upon identification of a correct clone cells are immediately available for template preparation, and only positive clones need be worked up. Clearly, one can also use the amplification product itself, for example in direct sequencing or recloning.

Protocol 10. Direct PCR screening of plaques and colonies

A. *To screen M13 clones as plaques:*

1. Prepare and label *n* 0.5-ml tubes, and (if rescue is required, e.g. to prepare sequencing template) *n* sterile culture tubes containing 1.5 ml 2 × TY medium.

2. Make up the following reaction mix, in bulk for (*n*+1) reactions:

   per reaction:
   - H_2O 14.9 μl
   - 10 × PCR buffer   2 μl
   - 5 mM dNTPs   1 μl
   - Primer 1 (10 pmol/μl)   1 μl
   - Primer 2 (10 pmol/μl)   1 μl
   - *Taq* polymerase   0.1 μl

   Aliquot 20 μl into each tube (conveniently performed using an Eppendorf multidispenser).

3. Gently stab the centre of a well-isolated plaque with a new toothpick, and swirl the toothpick briefly in the PCR reaction mix. Then place the toothpick into the appropriate culture tube if applicable. Repeat for each plaque.

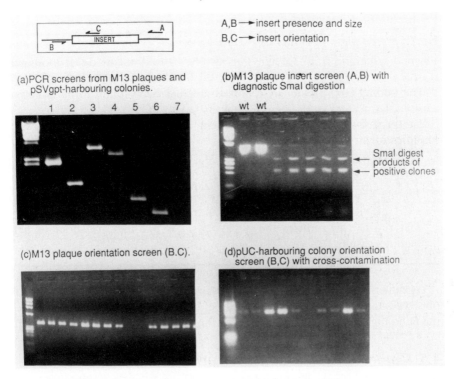

A,B→insert presence and size

B,C→insert orientation

(a)PCR screens from M13 plaques and pSVgpt-harbouring colonies.

(b)M13 plaque insert screen (A,B) with diagnostic SmaI digestion

Smal digest products of positive clones

(c)M13 plaque orientation screen (B.C).

(d)pUC-harbouring colony orientation screen (B,C) with cross-contamination

Figure 9. PCR screening. **(a)** shows amplifications of various inserts up to 3.3 kb in length from plasmid colonies [1, 2] and M13 plaques [3–5]. 6 and 7 are positive and negative M13 plaques in an orientation screen. The products of the **(A, B)** screen in **(b)** were analysed by *Sma*I digestion (5 units added directly to the PCR tube after cycling, and incubated at 25°C for 1 h), to screen for successful introduction of the site in a mutagenesis experiment. wt = wild type, the original starting clone.

4. Cover each reaction with paraffin as a single drop per tube from a Gilson P-1000 pipette equipped with a blue tip.

5. Amplify. For the M13 forward and reverse sequencing primers (17-mers) and a 1 kb insert, a suitable programme would be:

- 94°C, 2 min
- (94°C, 1 min; 55°C, 1 min; 72°C, 1 min) × 25 cycles
- 72°C, 5 min.

The choice of PCR conditions is discussed in Section 4.2.

6. If clones are to be grown up from the toothpicks, place the culture tubes shaking vigorously at 37°C (see Section 4.4).

7. Analyse 5 µl of the PCR product on an appropriate agarose gel.

Protocol 10. *Continued*

B. *To screen plasmid clones as bacterial colonies*

Plasmid screening from colonies can also be carried out by direct toothpick transfer.

The boiling step previously reported (17) has been found to be unnecessary.

The procedure is the same as screening M13 plaques, with the following modifications:

1. Hold the temperature at 94°C for 5 min to effect cell lysis before starting cycles.

2. Lightly touch the edge of the colony with the tip of the toothpick, then swirl into the reaction mix. Do not transfer a large scrape of colony material as this reduces the quality of PCR and can give poor or ambiguous results.

As shown in *Figure 9d*, occasional cross-contamination of negative clones is observed and in extreme cases can inhibit interpretation of results. Minimizing handling of toothpicks over the reaction tubes usually solves this problem.

4.1 Choice and availability of primers

Use a pair of primers flanking the insert (A and B in *Figure 9*) will yield a PCR product corresponding to the size of the insert. A single flanking primer and one hybridizing within the insert (B and C) will only give a PCR product in one orientation (the 'plasmid backbone' PCR product is not observed). The A and B type primers are usually available for any given vector system as forward and reverse sequencing primers, and often in practice a pre-existing internal sequencing primer for the insert can be used as primer C.

4.2 Selection of PCR conditions

The principles described for PCR from cloned DNA in Section 3.1.1 apply. Using 96-well plates (Techne FMW-1) and an appropriate heating block (Techne MW1) greatly simplifies manipulations.

4.3 Analysis of PCR screen products

In some cases it may not be possible to use either an (A,B) or (B,C) screen to distinguish a desired construct from a background of original clones; for example, after swapping two similarly-sized gene segments by 'sticky feet'-directed mutagenesis. In these cases digestion of the PCR screen product with a suitable restriction enzyme will often reveal correct clones (see *Figure 9b*). Most enzymes tested by the authors work at least adequately when simply added into the PCR reaction after cycling is complete; alternatively, use 5 µl of the PCR products in a separate digest.

4.4 Rescuing clones after setting up PCR screens

4.4.1 M13

If a plaque purification step is required, for example after mutagenesis, place the toothpick into the 1.5 ml medium and vortex to mix. Streak out this diluted phage suspension on a TYE plate and overlay with 3 ml top agar containing 100–200 µl of a suitable exponentially growing culture. Incubate the plates at 37°C overnight. To obtain sequencing template or RF-DNA from the 1.5 ml cultures, harvest by standard procedures (18) after shaking at 37°C for 6–7 h. Alternatively, innoculate a 1:100 dilution of a suitable bacterial overnight culture (rather than medium alone) with the toothpick, and grow for 4.5–5 h.

4.4.2 Plasmids

Grow clones from toothpicks in 1.5 ml medium containing suitable antibiotics for 8–16 h before harvesting. Superinfection by helper phage at an appropriate cell density can be used to rescue single-stranded phagemid genomes for sequencing (22). Overnight cultures in 200 µl medium in a 96-well plate (Nune) are very convenient for multiple samples.

5. Suppliers of enzymes and equipment

We have used native *Taq* polymerase supplied by Cetus. The Promega enzyme is a more economical alternative. All other enzymes were supplied by New England Biolabs unless otherwise indicated. PCR was performed using PHC-1 and PHC-2 programmable heating blocks manufactured by Techne.

Acknowledgements

We thank our colleagues for sharing protocols and advice, in particular Ermanno Gheradi, Andrew Griffiths, Robert Hawkins, and Sally Ward. This work was carried out in the laboratory of Greg Winter, whom we gratefully acknowledge for details of the VH/VK primer design and for encouragement.

References

1. Newton, C. R., Graham, A., Hepinstall, L. E., Powell, S. J., Summers, C., Kalsheker, N., Smith, J. C., and Markham, A. F. (1989). *Nucleic Acids Research*, **17**, 2503.
2. Feener, C. A., Koenig, M., and Kunkel, L. M. (1989). *Nature*, **338**, 509.

3. Wilks, A. F. (1989). *Technique*, **1**, 66.
4. Orlandi, R., Güssow, D. H., Jones, P. T., and Winter, G. (1989). *Proceedings of the National Academy of Sciences, USA*, **86**, 3833.
5. Ward, E. S., Güssow, D., Griffiths, A. D., Jones, P. T., and Winter, G. (1989). *Nature*, **341**, 544.
6. Kabat, E. A., Wu, T. T., Reid-Miller, M., and Gottesman, K. S. (1987). In *Sequences of proteins of immunological interest*. US Department of Health and Human Services, US Government Printing Office.
7. Kwok, S. and Higiuchi, R. (1989). *Nature*, **339**, 237.
8. Sarker, G. and Sommer, S. S. (1990). *Nature*, **343**, 27.
9. Hanish, J. and McClelland, M. (1988). *Gene Analysis Techniques*, **5**, 105.
10. Güssow, D., Ward, E. S., Griffiths, A. D., Jones, P. T., and Winter, G. (1990). *Cold Spring Harbor Symposia on Quantitative Biology*, **LIV**, 265–72.
11. Hemsley, A., Arnheim, N., Toney, M. D., Cortopassi, G., and Galas, D. J. (1989). *Nucleic Acids Research*, **17**, 6545.
12. Clackson, T. P. and Winter, G. (1989). *Nucleic Acids Research*, **17**, 10163.
13. Kunkel, T. A., Roberts, J. D., and Zakour, R. A. (1988). *Methods in enzymology*, Vol. 154 (ed. R. Wu), pp. 367. Academic Press, New York and London.
14. McClary, J. A., Witney, F., and Geisselsoder, J. (1989). *BioTechniques*, **7**, 282.
15. Higiuchi, R., Krummel, B., and Saiki, R. (1988). *Nucleic Acids Research*, **16**, 7351.
16. Horton, R. M., Hunt, H. D., Ho, S. N., Pullen, J. K., and Pease, L. R. (1989). *Gene*, **77**, 61.
17. Güssow, D. and Clackson, T. P. (1989). *Nucleic Acids Research* **17**, 4000.
18. Sambrook, J., Fritsch, E. F., and Maniatis, T. (1989). *Molecular cloning: A laboratory manual*, 2nd edn, pp. 4.26–4.32. Cold Spring Harbor Laboratory Press, Cold Spring Harbor, NY.
19. Ferre, F. and Garduno, F. (1989). *Nucleic Acids Research*, **17**, 2141.
20. Palmiter, R. D. (1974). *Biochemistry*, **13**, 3606.
21. Ullrich, A., Shine, J., Chingrin, J. *et al* (1977). *Science*, **196**, 1313.
22. Sambrook, J., Fritsch, E. F., and Maniatis, T. (1989). *Molecular cloning: A laboratory manual*, 2nd edn, pp. 4.44–4.48. Cold Spring Harbor Laboratory Press, Cold Spring Harbor, NY.

13

Semi-quantitative PCR for the analysis of gene expression

MARGARET J. DALLMAN and ANDREW C. G. PORTER

1. Introduction

The polymerase chain reaction (PCR) is an invaluable technique for the analysis of gene expression since mRNA reverse transcribed into cDNA may be subjected to conventional PCR amplification. This is particularly useful when analysing low abundance mRNAs or when limiting amounts of material are available (see Chapter 10). For certain situations there is clearly a need for an approach that will quantitate the amount of a particular mRNA in the initial sample.

Despite the sometimes unpredictable quantities of material obtained in PCR, two methods have been reported for the quantitation of mRNA. In both methods, an internal standard target sequence, which may be either DNA or RNA, is used as a control although clearly the use of RNA rather than DNA provides a control not only for the amplification, but also for cDNA synthesis. In the first method, a known quantity of the internal standard is used in the PCR against which the unknown or experimental nucleic acid may be compared following co-amplification. The internal standard may either have different (1), or identical (2) primer requirements to the experimental target DNA. The former approach has the advantage that two pairs of primers are used in the reaction so that the experimental and control targets do not compete for primers. However, this feature may also be seen as a disadvantage in that differences in primer efficiency between primer pairs may occur. The latter approach uses an internal standard which has the same primer requirements as the unknown mRNA but differs in product size or by the presence of restriction sites. However, this method has the problem of competition for primers between the standard and experimental target sequences and if either is present in excess it will influence the amount of amplification of the other. In fact, this competitive effect has been used advantageously for quantitative PCR (3). In this technique an internal standard with the same primer requirements as the experimental cDNA is used, but with a different product size. However, the amount of standard DNA is titrated against a constant amount of the experimental cDNA. The concentration of standard at which the amounts of amplification products from both targets is equivalent is taken to equal the starting concentration of the experimental cDNA. In both of these approaches the fact that the standard and experimental sequences are different may affect the point at which a plateau in PCR product is obtained and thus the accuracy of quantitation.

In many situations it may be sufficient and simpler to compare the relative amounts of a particular target sequence in different samples. We describe such an approach in this chapter. The essence of the approach is to sample product from the PCR at multiple points throughout the amplification process, thus ensuring analysis of product before a plateau is reached. Using this method, tenfold differences in the level of starting mRNA are easily detectable. A comparison of the level of cytokine mRNA in different cell populations has been performed using this approach. The use of primers to a control transcript, in this case to actin mRNA, shows that each sample has undergone equivalent reverse transcription to cDNA and equivalent amplification in the PCR. Since the control sequence is not being used to define absolute levels of mRNA, differences in primer efficiencies between the control and test primers are not important. This approach has the advantage of simplicity and may readily be applied to any previously defined PCR conditions making it easily accessible to the analysis of different target sequences.

2. cDNA synthesis

Total cytoplasmic RNA is prepared by standard protocols (4). For a semi-quantitative analysis the use of small amounts (e.g. 1 μg) of total RNA in the cDNA synthesis give the best results (*Protocol 1*). Oligo-dT is used as a primer, so that the same cDNA preparation may be used for the analysis of multiple mRNAs. A small sample of the cDNA is trace radiolabelled during synthesis and subsequently analysed by alkaline agarose gel electrophoresis to ensure that all samples have undergone equivalent reverse transcription. Gel electrophoresis enables an estimate to be made of both the size and quantity of cDNA. The average size of cDNA should be ~1–2 kb but should be easily visible up to about 6–7 kb. A more accurate estimate of the efficiency of cDNA synthesis may be obtained by measurement of TCA precipitable counts. It is preferable that the cDNAs to be compared in PCR be synthesized in parallel using the same reaction mix.

Protocol 1. cDNA synthesis

1. Set two water baths to 60°C and 37°C.
2. Dry down 5 μCi [α-^{32}P] dCTP in a microcentrifuge tube.
3. Resuspend 1 μg total RNA in 19 μl double deionized H$_2$O (ddH$_2$O) in a microcentrifuge tube and add 4 μl oligo-dT(1 mg/ml).

4. Denature RNA by incubation at 60°C for 5 min. NB: for certain mRNAs with a high degree of tertiary structure a higher temperature of 80°C should be used for this denaturation step.

5. Allow oligo-dT to anneal to RNA by chilling on ice for 3 min.

6. Turn 60°C water bath to 70°C.

7. Add: ● 0.5 µl RNAsin (5 units), ● 4 µl 10 mM dNTPs ● 2 µl reverse transcriptase (RT-Moloney murine leukaemia virus, 200 units/µl (Gibco-BRL), ● 8 µl RT buffer (provided as a 5 × concentrated solution with RT), and mix. For preparation of multiple cDNAs pre-mix these reagents in one batch and divide between RNA samples.

8. Remove 5 µl of each mix to the microcentrifuge tubes containing the ^{32}P-dCTP and resuspend well.

9. Incubate all tubes at 37°C for 40 min. NB: the temperature of this incubation may vary if different RT is used.

10. Add 2 µl RT to main mix (i.e. the one without the ^{32}P dCTP).

11. Incubate at 37°C for 40 min.

12. Heat inactivate RT by incubation at 70°C for 10 min.

13. Add 60 µl ddH$_2$O to main samples. Store samples at −20°C until use.

14. Check cDNA synthesis by alkaline agarose gel electrophoresis of 1–2 µl of the ^{32}P-labelled cDNA (4) using the ^{35}S-labelled markers to assess size.[a] Dry down gel before exposing to film.

---

[a] Markers are prepared by adding 10 µg DNA (either 1 kb ladder (Gibco-BRL) or *Hind*III digested λ) to 10 µl buffer (50 mM NaCl, 10 mM Tris–Cl pH 7.5, 10 mM MgCl$_2$, 1 mM dithiothreitol); 10 µl each of 1 mM dTTP, dCTP, dGTP; 50 µCi [α-^{35}S] dATP; 2 units Klenow, and make up to a total volume of 100 µl. Incubate 37°C for 60 min and separate on a spun column (4). Run 3–5 µl on the gel.

---

3. Polymerase chain reaction

For mRNAs of average or low abundance, PCR may be performed using 5 µl of the cDNA prepared in *Protocol 1*. One batch of cDNA may therefore be used in the analysis of multiple mRNAs. The reaction mix is sampled at multiple time-points during PCR for subsequent analysis. For low abundance mRNAs sample from 20 cycles and for high abundance from 10–15 cycles. Sampling every 5 cycles easily allows discrimination of tenfold or greater differences in the original level of target sequence. The particular conditions (i.e. MgCl$_2$ concentration and annealing temperature) should be adjusted for each primer pair. *Protocol 2* gives conditions for amplification from the actin primer pair which have the sequences 5′ATGGATGACGATATCGCTG 3′;

5'ATGAGGTAGTCTGTCAGGT 3' and which generate a product of 568 bp. Always use a no cDNA control.

Protocol 2. Polymerase chain reaction

Perform steps 1 to 3 then put reagents away before handling the cDNA to avoid the risk of contaminating reagents.

1. Prepare the amplification mix by combining the following for each cDNA sample: ● 5 μl 1 M KCl; ● 1 μl 1 M Tris–HCl pH 8.3; ●2.5 μl 0.1 M MgCl$_2$; ● 5 μl 0.2% gelatin; ● 4 μl 20 mM dNTPs; ● 2 μl each of 5' and 3' primers (50 μM); ● 0.5 μl *Taq* polymerase (5 units/μl). Make up to 50 μl with ddH$_2$O. Prepare sufficient mix for one reaction more than the number of samples to allow for pipetting errors.

2. Aliquot 50 μl amplification mix into 0.5 ml microfuge tubes.

3. Add ddH$_2$O to 100 μl less the volume of cDNA to be added (i.e. normally 45 μl ddH$_2$O when 5 μl cDNA is being used).

4. Add 5 μl cDNA.

5. Add 50 μl mineral oil to prevent evaporation from samples.

6. Cycle at 94°C, 1 min; 55°C, 2 min; 72°C, 1 min. The annealing temperature, in this case 55°C, will vary with the primers used.

7. Remove 15 μl samples every 5 cycles starting after 10–20 cycles. This is best performed at the end of a cycle, i.e. at the end of the 72°C extension phase. Remove all the tubes together and sample using a Pipetman (Gibson). Take a small amount of air into the Pipetman tip, push tip through the oil and blow out the air before aspirating 15 μl of the sample. This avoids mixing of the sample with oil. Samples may be conveniently stored in 96-well, round-bottomed plates. These plates are also a convenient vessel in which to prepare samples for dot-blotting.

---

4. Analysis of PCR products

Analysis of PCR samples is performed either by agarose gel electrophoresis followed by Southern blotting (4) or by dot-blotting and then hybridized to an internal olignucleotide that lies within the target sequence between the two original primers. It is always essential to establish by gel electrophoresis that under the conditions used the PCR results in a single product of the expected size. Confirmation that the product is specific may be obtained by Southern blotting the DNA and probing with the internal oligo. For analysis of products greater than 0.4 kb in size, an ordinary 1% agarose gel is fine. For smaller fragments use a 3% NuSieve agarose (FMC BioProducts, supplied by

ICN Biochemicals) and 1% normal agarose. Dot-blotting PCR products and probing with an internal oligo (see *Protocol 3*) provides a convenient method for the analysis of large numbers of samples. Use of internal oligos as probes confers an additional level of specificity in the reaction as only the specific PCR product will bind the oligo under the correct stringency of hybridization and washing. The sequence of the internal oligo for the actin gene is 5′AGCAAGAGAGGTATCCT 3′.

Protocol 3. Dot-blot analysis of PCR samples.

1. Make volume of PCR samples up to $100\,\mu l$ with TE ($10\,mM$ Tris–HCl pH 8.0, $0.1\,mM$ EDTA).

2. Set up dot-blot manifold with two layers pre-wetted Whatman 3MM paper underneath one layer pre-wetted nitrocellulose/nylon membrane.[a] Pre-wash filter with $200\,\mu l$ 1 M NH_4OAc.

3. Add $6\,\mu l$ 0.5 M EDTA to each sample then to denature DNA, add $8\,\mu l$ 6 N NaOH and incubate on ice for 10 min.

4. Neutralize by adding $115\,\mu l$ 2 M NH_4OAc.

5. Load samples into dot-blot manifold and wash through with $200\,\mu l$ 1 M NH_4OAc.

6. Remove filter and soak for 15 sec in $6 \times$ SSC.

7. Air-dry filter and bake 80°C for 60 min.

8. Pre-hybridize for 2–4 h at 42°C with agitation in pre-hybridization solution: 1.5 ml $20 \times$ SSC; 1.0 ml $50 \times$ Denhardt's solution; 0.1 ml $20\,mg/ml$ tRNA; 0.25 ml 10% SDS; 2.25 ml ddH_2O.

9. Replace pre-hybridization solution with hybridization solution (0.75 ml $20 \times$ SSC; $25\,\mu l$ 10% SDS; 1.75 ml ddH_2O) containing ^{32}P end-labelled internal oligo (about $4–8 \times 10^6$ c.p.m/blot).

10. Incubate at 42°C overnight with agitation.

11. Wash blot in 200 ml $6 \times$ SSC; 0.1% SDS for 5 min at r.t. followed by a second wash for 5 min in 200 ml fresh wash solution at the T_m of the oligo (i.e. [$4 \times$ CG content] + [$2 \times$ AT content]). T_m for actin internal oligo = 50°C.

12. Expose to X-ray film. Usually dot-blots of PCR products only require a few minutes to 3 h exposure. Multiple exposure lengths are often a useful aid to analysis. Alternatively, analyse using PhosphoImaging or cut out dots and analyse by β-scintillation counting (see Section 5.5).

---

[a] To save membrane only cut a piece sufficient for the number of samples to be blotted, but cover all holes with 3MM paper. Nitrocellulose membranes give cleaner and more reproducible results.

---

5. Experimental examples using this approach

5.1 Single reaction product

PCR using IL-2 or actin primers results in single reaction products (see *Figure 1*). If a single product is not obtained in PCR, then the approach described in this chapter for semi-quantitative PCR is invalidated, because conditions in the amplification may become limiting before a maximal amount of specific

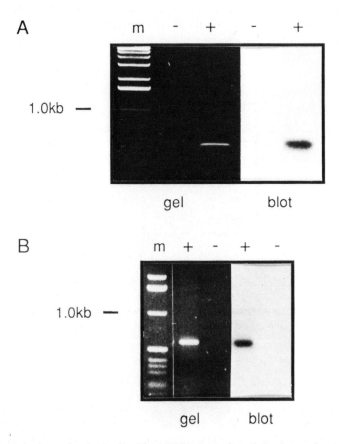

Figure 1. A single product is obtained from PCR using actin or IL-2 primers. 1 μg total RNA obtained from a rat renal allograft 5 days after transplantation was used to synthesize cDNA. 5 μl of the cDNA was used in PCR with the following primers: 5'AACAGCGCACCCACTTCAA3' and 5'TTGAGATGATGCTTTGACA3' (to give an expected product size of 400 bp), using 1.5 mM MgCl₂, 60 °C annealing temperature and 45 cycles **(A)**, or with actin primers as described in *Protocol 2* **(B)**. 15 μl product was electrophoresed on a 1% agarose gel and blotted to nitrocellulose and probed with the appropriate ³²P-labelled internal oligo. The internal oligo for IL-2 has the sequence 5'CACTGAAGATGTTTC3' and for actin as described in *Protocol 3*. Blots were washed at the T_m of the oligo, i.e. 50 °C and 42 °C for actin and IL-2 respectively. + = with cDNA; −= without cDNA; m = marker track (1 kb ladder).

product is obtained. For this reason it is important to check for a single PCR product by gel electrophoresis in ethidium bromide containing agarose, followed by Southern blotting of the gel.

5.2 cDNA dilutions

PCR results using tenfold dilutions of a single cDNA sample are shown in *Figure 2*. In this experiment, 0.1 µl, 1.0 µl, or 10 µl of cDNA from a synthesis using 10 µg total RNA were amplified using IL-2 primers. The difference in level of PCR product is easily detectable following dot-blotting of samples. The no cDNA control gives no product.

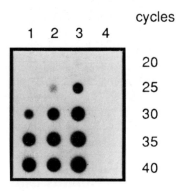

Figure 2. PCR using tenfold dilutions of cDNA. 0.1 µl **(column 1)**, 1.0 µl **(column 2)**, 10 µl **(column 3)** of a cDNA synthesis using 10 µg RNA or no cDNA **(column 4)** were used in PCR with IL-2 primers. 15 µl samples were removed at 20, 25, 30, 35, and 40 cycles, dot-blotted to nitrocellulose membrane and probed using the IL-2 internal oligo. Exposure to film was 30 min at room temp. Source of RNA and conditions for PCR and blot-washing are described in *Figure 1*.

5.3 RNA dilutions

cDNA synthesis and PCR using tenfold dilutions of a single RNA preparation are shown in *Figure 3*. In this experiment, the same RNA preparation was used in different cDNA synthesis reactions but using tenfold dilutions of the RNA (i.e. 0.1, 1.0, 10, and 100 µg RNA). This controls for both the cDNA synthesis and amplification reactions. Again, the tenfold difference in target sequence is easily detectable following PCR using IL-2 primers. Control experiments where the smaller quantities of RNA were made up to 100 µg with either tRNA or linear polyacrylamide showed that the reactions are not affected by using smaller amounts of material in the cDNA synthesis. It is very useful to include this type of control in each experiment that is performed.

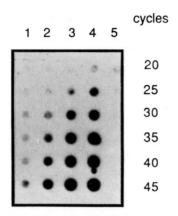

Figure 3. cDNA synthesis and PCR using tenfold dilutions of RNA. 0.1 µg (**column 1**), 1.0 µg (**column 2**), 10 µg (**column 3**), 100 µg (**column 4**) or no (**column 5**) RNA was used in cDNA synthesis. 5 µl of each cDNA was then amplified under identical conditions using IL-2 primers. 15 µl samples were removed at 20, 25, 30, 35, 40, and 45 cycles, blotted on to nitrocellulose membrane and probed using ^{32}P-labelled internal IL-2 oligo. Exposure to film was 3 h at room temp. Source of RNA and conditions for PCR and blot-washing are described in *Figure 1*.

5.4 Comparison of IL-2 mRNA in syngeneic and allogeneic rat renal transplants using comparative PCR

Different samples of RNA were compared for their content of IL-2 mRNA. Syngeneic transplants in which little immune response occurs contain virtually undetectable levels of IL-2 mRNA. However, allogeneic transplants, in which a strong immune response occurs, contain increasing amounts of IL-2 mRNA with time after transplantation (*Figure 4a*). This experiment is controlled using PCR with actin primers and in all cases similar levels of product are obtained (*Figure 4b*). The difference between corresponding dots from syngeneic and allogeneic samples is striking, reproducible and clearly greater than the differences observed between dots representing tenfold, and in most cases 100-fold dilutions in control samples. We can therefore conclude that IL-2 expression in the grafts of rejecting animals is at least 10–100 times greater than that in non-rejected grafts.

5.5 Measurement of signal intensities on dot-blots

Due to the non-linearity of X-ray film, dots or bands on a blot that contain large, but different amounts of PCR product will appear of the same intensity even after very short exposure to film. One way of overcoming this problem, is to carry out multiple short exposures and to analyse the film using densitometry. As an alternative, PhosphoImaging may be used. *Figure 5* shows

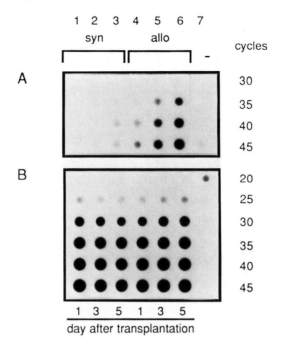

Figure 4. Comparison of IL-2 and actin mRNA in syngeneic and allogeneic rat renal transplants using comparative PCR. 1 μg total RNA was isolated from rat renal allografts in either the DA->DA (syngeneic, non-rejecting **columns 1–3**) or DA->PVG (allogeneic, rejecting, **columns 4–6**) strain combinations, 1 (**columns 1, 4**), 3 (**columns 2, 5**) or 5 (**columns 4, 6**) days after transplantation and converted to cDNA. 5 μl of the cDNA was used in PCR with either IL-2 **(A)** or actin **(B)** primers. 15 μl samples were removed at 20, 25, 30, 35, 40, and 45 cycles, dot-blotted to nitrocellulose membranes and probed with the appropriate ^{32}P-labelled internal oligo. No cDNA control column 7. Exposure to film, 75 min, room temperature. Conditions for PCR and blot-washing are discribed in *Figure 1*.

that although several of the dots on this particular blot appear to be of the same intensity after only 30 min exposure of film at room temperature, PhosphoImaging clearly can distinguish an increasing amount of bound probe. Similar numerical data may be obtained by simply analysing the dots using β scintillation counting.

6. Summary

An easy and rapid method for a comparative and semi-quantitative analysis of PCR products has been described. This approach is particularly useful in the analysis of non-constitutively expressed, tissue-specific genes, whose mRNA is present in low abundance—in this case the rat IL-2 gene. For many purposes of this type of analysis will be sufficient. The approach does suffer from

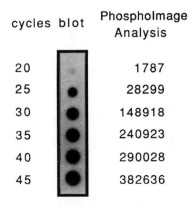

cycles	blot	PhosphoImage Analysis
20		1787
25		28299
30		148918
35		240923
40		290028
45		382636

Figure 5. Analysis of blots by PhosphoImaging. cDNA synthesis, PCR and blotting were performed as in *Figure 1* using 10 μl cDNA from a synthesis using 10 μg RNA. The blot was exposed to film for 30 min at room temp. PhosphoImaging was performed using the Molecular Dynamics PhosphoImager with a 5-min expsoure. Figures obtained are shown in the right-hand column.

lack of an internal standard although, as previously discussed, the inclusion of such controls is not without problems. To overcome this limitation, analysis of samples is always repeated and has proved to be reproducible. Further, by using control primers we always establish that (a) conversion of RNAs to cDNA, and (b) PCR amplification are equivalent between samples.

As exemplified in *Figure 4*, this technique can be very useful for observing strong induction of low abundance mRNAs in situations where Northern blots are insufficiently sensitive and where limiting amounts of material prohibit the use of RNase protection analysis.

References

1. Chelly, J., Kaplan, J.-C., Maire, P., Gautron, S., and Kahn, A. (1988). *Nature*, **333**, 858.
2. Wang, A. M., Doyle, M. V., and Mark, D. F. (1989). *Proceedings of the National Academy of Sciences, USA*, **86**, 9717.
3. Gilliland, G., Perrin, S., Blanchard, K., and Bunn, H. F. (1990). *Proceedings of the National Academy of Sciences, USA*, **87**, 2725.
4. Sambrook, J., Fritsch, E. F., and Maniatis, T. (ed.) (1989). *Molecular cloning: A laboratory manual*, 2nd edn. Cold Spring Harbor Laboratory Press, Cold Spring Harbor, NY.

14

The fidelity of DNA polymerases used in the polymerase chain reactions

KRISTIN A. ECKERT and THOMAS A. KUNKEL

1. Introduction

Purified DNA polymerases are one of the primary tools for molecular biology. Yet choosing the most appropriate DNA polymerase for any particular application requires an understanding of the substantial biochemical differences between the available enzymes. One aspect of interest for PCR is the fidelity of DNA polymerization, i.e. the number of errors produced per nucleotide synthesized. We begin this chapter by discussing the enzymology of error discrimination by DNA polymerases during DNA synthesis *in vitro*. We then describe the fidelity of several prokaryotic DNA polymerases using a simple assay for scoring DNA polymerase errors. Next we examine the parameters that affect the fidelity of the polymerase most often used for PCR, the *Taq* DNA polymerase isolated from *Thermus aquaticus*. Finally, we describe the implications of these fidelity studies for PCR, and end with some general recommendations for performing high fidelity amplification reactions.

2. Steps in polymerase error discrimination

DNA synthesis by DNA polymerases is a highly ordered and complex molecular process, during which a deoxynucleotide triphosphate (dNTP) substrate is added to a free 3′-hydroxyl of a DNA primer-template (1, 2). At least three opportunities exist for the polymerase to discriminate against errors during synthesis. For simplicity, these steps are described in terms of single-base substitution errors (*Figure 1*). The overall fidelity of a polymerase reaction depends upon the relative error rates of each of the three steps. Discrimination at each step can be expected to vary according to the individual polymerase, local DNA sequence, and base mispair under consideration.

The first step in error discrimination is the ability of a DNA polymerase to distinguish among incoming dNTP substrates and to incorporate the correct

225

A. Nucleotide insertion

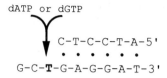

```
dATP or dGTP

       C-T-C-C-T-A-5'
       • • • • • •
   G-C-T-G-A-G-G-A-T-3'
```

B. Mispair extension

```
     A-C-T-C-C-T-A-5'
     • • • • • • •
   G-C-T-G-A-G-G-A-T-3'
```

or

```
     G-C-T-C-C-T-A-5'
     • • • • • •
   G-C-T-G-A-G-G-A-T-3'
```

C. Exonucleolytic proofreading

```
     C-T-C-C-T-A-5'
     • • • • • •        + dGMP
   G-C-T-G-A-G-G-A-T-3'
```

Figure 1. Steps in polymerase error discrimination.

rather than an incorrect nucleotide on to a DNA primer-template (*Insertion step, Figure 1A*). During DNA synthesis, there are twelve potential base mispairs by composition and symmetry (template base versus incoming deoxynucleoside triphosphate). Together with the effects of neighbouring template sequence, this creates a diversity of possible molecular structures among which any polymerase must discriminate. Polymerases differ in their abilities to distinguish among these various structures, and ten- to 100-fold variations in base substitution error rates at this step are not unusual (3, 4). (For a more detailed discussion of polymerase discrimination at the insertion step, see refs 5 and 6.) Base selectivity at the insertion step is also affected by the relative concentrations of each of the four deoxynucleotide triphosphates. Deoxynucleotide pool imbalances can be mutagenic or antimutagenic depending on the error being considered, reflecting the relative probability that a polymerase will incorporate an incorrect versus a correct dNTP on to a nascent DNA chain.

The second step is the selective ability of a polymerase to continue synthesis on a correctly paired primer-template versus a primer-template containing a terminal mispair (*Extension step, Figure 1B*). As with the insertion step, the rates of extension for any particular polymerase can vary widely according to the mispair symmetry and composition and the local template sequence (4, 7). The probability of fixing any given mispair by extension synthesis is increased as the concentration of the next correct nucleotide(s) increases, as this will drive the enzyme reaction in the direction of DNA polymerization.

Thirdly, some, but not all, DNA polymerases have an associated $3' \rightarrow 5'$ exonuclease activity (for review, see ref. 8), which selectively removes mis-incorporated nucleotides to regenerate a correctly base-paired primer-terminus (*Proofreading step, Figure 1C*). Variations in proofreading efficiency reflect the balance between the competing processes of polymerization and excision. Proofreading can be diminished by high concentrations of dNTP substrates, which increase the rate of extension synthesis from mispaired termini. The addition of nucleotide monophosphates also reduces proofreading by binding to the exonuclease active site and preventing excision. As with the other steps in error discrimination, the efficiency of proofreading depends upon the enzyme and the precise molecular nature of the error under consideration.

3. Polymerase fidelity *in vitro*

The error rate of DNA synthesis *in vitro* is not a constant, but rather depends on the polymerase, the DNA sequence, the error under consideration, and the reaction conditions. In this section, we focus on the fidelity of four commonly used DNA polymerases: T4 DNA polymerase, native T7 DNA polymerase, the large (Klenow) fragment of *Escherichia coli* DNA polymerase I, and *Taq* DNA polymerase, all of which can be used in PCR amplifications (9).

3.1 The M13mp2 fidelity assays

Several different methods are available to detect errors made by DNA polymerases during *in vitro* DNA synthesis (5, 6, 9, 10). The M13mp2 forward mutation assay provides a broad description of the possible DNA polymerase errors that can occur during *in vitro* DNA synthesis using a natural DNA template (11). This assay scores for loss of a non-essential gene function, *lacZ* α-complementation of β-galactosidase activity, and therefore detects a wide variety of mutations. All twelve possible base substitution errors at 114 different sites can be scored within the 258-base *lacZ* α target sequence, along with frameshift mutations, deletions and more complex errors. In this assay (*Figure 2*) DNA polymerase errors are measured as a result of a single round of *in vitro* DNA synthesis to fill a 390 base gap located opposite the wild-type M13mp2 DNA sequence. Accurate *in vitro* DNA synthesis results in dark blue M13mp2 plaques after transfection of *E. coli* with the reaction products, while polymerase errors during gap-filling synthesis result in M13mp2 mutant plaques having light blue or no colour due to decreased α-complementation of β-galactosidase activity in the infected host cell.

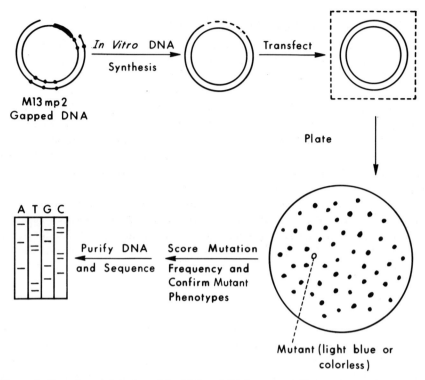

Figure 2. Experimental design of the M13mp2 fidelity assays. The *lacZα* mutational target is represented by the bold line within the M13mp2 gapped DNA molecule. *In vitro* gap-filling synthesis by DNA polymerases occurs from right to left. The resulting full length RFII molecules are used to transfect competent *E. coli* cells (dashed box), and polymerase errors are monitored as changes in M13mp2 plaque colour.

Specific types of polymerase errors can be examined using M13mp2 reversion assays. In these assays, the strategy is the same as for the forward assay except that the DNA templates are made using a defined M13mp2 *lacZ* α mutant sequence which has a colourless plaque phenotype. In these cases, DNA polymerase errors during *in vitro* synthesis are detected as blue 'revertant' phage plaques, while accurate polymerization yields colourless plaques. Base substitution errors can be monitored in the opal codon reversion assay which detects eight possible single-base-substitution errors (12), while the frameshift reversion assay measures the loss of a single base (4).

In all M13mp2 assays, polymerase errors are quantitated from the mutant frequency, defined as the proportion of mutant plaques to the total number of plaques scored. The molecular nature of the error can be determined by DNA sequence analysis of the mutants. Using data generated by these *in vitro* fidelity assays the error rate of a polymerase, during a single cycle of DNA synthesis, can be precisely calculated for each specific class of polymerase error.

Kristin A. Eckert and Thomas A. Kunkel

3.2 General DNA polymerase fidelity

Although all DNA polymerases synthesize DNA by the same basic reaction mechanism they differ in several biochemical properties (1), including fidelity. The best means of achieving high fidelity DNA synthesis is to use a DNA polymerase which contains a 3′→5′ proofreading exonuclease to remove nucleotides that have been misinserted during polymerization. All the proof-reading-proficient polymerases listed in *Table 1* contain the polymerase and 3′→5′ exonuclease activities within the same polypeptide, but these enzymes differ in the relative strengths of their proofreading activities. Both the T4 and native T7 DNA polymerases contain highly active proofreading activities and are very accurate for base substitution and one base frameshift mutations, even when high concentrations of dNTP substrates are present during synthesis (*Table 1*). The Klenow polymerase also contains a 3′→5′ exonuclease, but the proofreading activity is much weaker than that associated with either the T4 or native T7 DNA polymerase. Thus, the Klenow polymerase is accurate during DNA synthesis using 1 μM dNTPs, a concentration that allows the exonuclease to proofread errors. However, at higher dNTP concentrations, such as those typically used in PCR, polymerization is favoured over exonucleolytic removal of most errors, resulting in a significantly lower fidelity (*Table 1*).

The commercially available thermostable DNA polymerases do not contain 3′→5′ proofreading exonuclease activities (ref. 13 for *Taq* and unpublished observations for *Tth*). When the fidelities of various polymerases are compared using similar reaction conditions, the *Taq* and *Tth* DNA polymerases are less accurate than are the T4, native T7 or Klenow polymerases, consistent with *Taq* and *Tth* being proofreading-deficient (*Table 1*).

3.3 Variables that influence *Taq* polymerase fidelity

Although in many circumstances the high fidelity of proofreading-proficient DNA polymerases may be desirable, there are several disadvantages to using these enzymes as reagents in PCR. The T4, native T7 and Klenow DNA polymerases are heat-labile and operate with an optimal synthetic rate at 37°C. For reactions performed in this temperature range there is an increased probability of non-specific hybridization of primers to other regions of the DNA to be amplified. Also, at the 37°C catalytic temperature, formation of template DNA secondary structures may provide a barrier to DNA elongation and inhibit synthesis through the entire target sequence. In contrast, DNA polymerases isolated from thermostable bacteria are heat-stable, permitting DNA synthesis at higher temperatures (70–80°C). This not only reduces DNA secondary structure formation and the probability of aberrant priming (14), but also permits a single addition of polymerase at the beginning of the amplification process. Since thermostable polymerases are useful reagents for PCR, in this section we focus in more detail on variables that affect *Taq* polymerase fidelity.

Table 1. Fidelity of DNA polymerases used in PCR

DNA polymerase	3'→5' Exonuclease activity	Mutant frequency $\times 10^{4a}$	Error rate per nucleotide polymerized	
			Base substitution	Frameshift
T4	+	8	1/80 000	1/260 000
Native T7	+	14	1/53 000	1/190 000
Klenow (1 µM dNTPs)	+	11	1/170 000	≤1/1 500 000
Klenow	±	37	1/27 000	1/140 000
Taq	−	110	1/9 000	1/41 000
Tth	−	130	NS	NS

M13mp2 forward mutation assays were performed using 10 mM $MgCl_2$ and 1 mM dNTPs (except where noted). The T4, native T7 and Klenow polymerase reactions were buffered with 20 mM Hepes (pH 7.8) and 2 mM DTT, and incubated at 37°C. The *Taq* and *Tth* polymerase reactions contained 20 mM Tris–HCl (pH 7.2 at 70°C) and 50 mM KCl and were incubated at 70°C. Error rates per nucleotide were calculated as described (13) after DNA sequence analysis of 36 (T4) or 13 (native T7) mutants. The data for Klenow polymerase are from refs 4 and 13, and the data for *Taq* are from ref. 13. NS = no sequence analysis of mutants was performed, so error rates for base-substitutions and frameshifts could not be calculated.

[a] A background mutant frequency of 7×10^{-4} has been subtracted to correct for mutants not arising from polymerase errors during synthesis.

3.3.1 dNTP concentration

Both the relative and the absolute concentrations of the dNTP substrates are critically important in the production of base substitution errors by DNA polymerases. The effect of dNTP concentration on *Taq* polymerase fidelity is shown in *Table 2*. In the M13mp2 fidelity assays, the base substitution error rate improved twofold while the frameshift error rate improved five-fold as the dNTP concentration was lowered from 1 mM to 1 µM (*Table 2*). Similar effects on fidelity over the same dNTP concentration range were observed for a 3'→5' exonuclease-deficient form of the Klenow polymerase (4).

Two factors account for the observed modest (twofold) increase in base substitution fidelity over a 1000-fold dNTP concentration range. First, the reaction times at low dNTP concentrations were increased to achieve the amount of DNA synthesis required for the fidelity assays (*Table 2*). This fact could minimize error discrimination at the extension step by allowing additional time for the polymerase to extend mispaired primer-termini. Second, the majority of errors made by both the *Taq* polymerase (ref.13, and see below) and the exonuclease-deficient Klenow polymerase (4) result from T·dGTP mispairs. With the exonuclease-deficient Klenow polymerase as well as with other DNA polymerases, this mispair is both the most easily generated and the most easily extended of the twelve possible mispairs (3, 4, 7). Thus, the improvement in the final error rate of a DNA synthesis reaction may be limited by the least accurate reactions that occur, i.e., formation and fixation of T·dGTP mispairs.

Table 2. *Taq* polymerase fidelity versus dNTP concentration

Reaction conditions		Base substitution assay		Frameshift assay	
[dNTP] (μM)	Time (min)	Frequency $\times 10^6$	Error rate (per nt)	Frequency $\times 10^5$	Error rate (per nt)[a]
1000	5	330 (110)[b]	1/5500	15	1/22 000
100	10	340 (37)	1/5400	11	1/31 000
10	30	240 (83)	1/7600	6.9	1/54 000
1	120	160 (37)	1/12 000	4.2	1/110 000

DNA synthesis reactions were carried out using 20 mM Tris–HCl (pH 7.2 at 70°C), 50 mM KCl, 10 mM $MgCl_2$, and each of four dNTP at the indicated concentration. Gapped DNA substrate and *Taq* polymerase were present in a 1:1 molar ratio (2.8 units of *Taq*, 500 ng of gapped DNA). The mixtures were pre-warmed to 70°C for 5 min, enzyme was added, and the reactions allowed to continue at 70°C for the additional time indicated.

[a] Error rates were calculated assuming all mutations occurred within the run of five consecutive thymine bases. However, the phenotypes of all revertants were not confirmed by plaque purification; thus, the reported error rate may overestimate the true value.

[b] Values are mean for at least three independent experiments with the standard deviations given in parentheses.

High concentrations of dNTP substrates generally will increase the polymerase error rate by driving the enzymatic reaction in the direction of DNA synthesis, thereby decreasing the amount of error discrimination at the extension step. Conversely, low dNTP concentrations will tend to increase fidelity by influencing the rate at which the polymerase extends mispaired primer-termini. If the polymerase used in PCR lacks an exonuclease to remove terminal mispairs, such as the *Taq* polymerase, unextended errors will be lost because they do not yield full-length DNA products for further amplification. This type of error discrimination at low dNTP concentrations forms the basis for allele specific amplification in PCR (15).

Deoxynucleotide pool imbalances also modify the polymerase error rate by influencing nucleotide selectivity at the insertion step. *Table 3* illustrates the production of base substitution mutations by *Taq* polymerase in response to dNTP pool imbalances. In the first experiment, the pool bias targets misincorporation errors opposite the template T of the TGA opal codon by decreasing the concentration of correct dATP relative to incorrect dGTP, dTTP, and dCTP. This imbalance results in an increase in the base substitution error rate for T·dGTP, T·dCTP, and/or T·dTTP errors, approximately in proportion to the extent of the substrate imbalance (*Table 3A*). In the second experiment, the base substitution error rate was increased by raising the concentration of dGTP relative to the other three nucleotides. Here, errors result from misincorporation of dGTP opposite any of the three template nucleotides of the TGA codon, i.e., from T·dGTP, G·dGTP, and/or A·dGTP mispairs. Again,

Table 3. Effect of dNTP pool imbalances on the fidelity of *Taq* polymerase

[dNTP] (μ M)	Pool bias	Base substitution frequency $\times 10^6$	Error rate
A. 1000	none	290	1/2400 [a]
	5 × low A	920	1/650
	10 × low A	2100	1/290
B. 10	none	150	1/21 000 [b]
	5 × high G	1700	1/1100
	10 × high G	3300	1/550

The base substitution reversion frequency was followed as a function of two types of pool imbalances. (A) The concentration of dATP was lowered from 1 mM (no pool bias) to 200 μM (5 × low A) or 100 μM (10 × low A) while the concentrations of dGTP, dTTP and dCTP were held at a constant concentration of 1 mM. (B) The concentration of dATP, dCTP, and dTTP were held at a constant concentration of 10 μM while the concentration of dGTP was increased from 10 μM (no pool bias) to 50 μM (5 × high G) or 100 μM (10 × high G). In addition to the indicated dNTPs, each reaction contained 20 mM Tris–HCl (pH 7.2 at 70°C), 50 mM KCl, 10 mM MgCl$_2$, 250 ng gapped DNA substrate, and 1.4 units of *Taq* polymerase. Reactions were incubated at 70°C for 5 (A) or 30 (B) min.

[a] The error rate for the unbiased reaction was calculated for first position changes only. As determined by DNA sequence analysis of 26 mutants, first position mutants account for 85% of the total.
[b] The error rates for the unbiased reaction was calculated for errors resulting from misincorporation of dGTP at all three positions which account for 58% of the total.

the pool imbalance substantially increased the error rate of *Taq* polymerase (*Table 3B*). These results demonstrate that maximum fidelity in PCR will be achieved by performing reactions with equal concentrations of all four dNTP substrates, and that specific errors can be produced by employing selective substrate imbalances.

3.3.2 Free MgCl$_2$ concentration

The fidelity of *Taq* polymerase can be improved by decreasing the magnesium chloride concentration relative to the total concentration of dNTPs present in the reaction (16). *Figure 3* shows the response of *Taq* polymerase in base-substitution and frameshift reversion assays as the total concentration of MgCl$_2$ in the reaction is increased at a constant dNTP concentration. In both M13mp2 assays, fidelity was highest at a MgCl$_2$ concentration that was equimolar to the total concentration of dNTP substrates present in the reaction. Comparing the mutagenic response of *Taq* polymerase to MgCl$_2$ at three different deoxynucleotide triphosphate concentrations, we conclude that high fidelity DNA synthesis correlates with the level of free divalent cation present in the reaction. This is shown in *Figure 4*, where the base substitution mutant frequency is plotted as a function of the concentration of

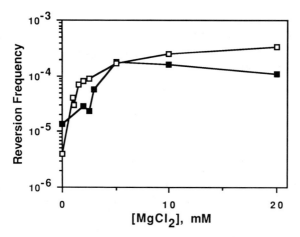

Figure 3. Fidelity of *Taq* polymerase versus magnesium chloride concentration. The mutagenic response of *Taq* polymerase to increasing MgCl$_2$ concentration is shown in the M13mp2 base-substitution (□) and framshift (■) reversion assays. Polymerase reactions were performed using 20 mM Tris–HCl (pH 7.2 at 70 °C) and 50 mM KCl, varying the [MgCl$_2$] as indicated (16). The deoxynucleotide triphosphate concentrations were 250 μM each dNTP (1 mM total) in the base substitution assay and 500 μM each dNTP (2 mM total) in the frameshift assay. Synthesis reactions at equimolar MgCl$_2$ and dNTP concentrations were incubated for 30 min at 70 °C. The data points for 'O' MgCl$_2$ indicate the background frequencies for the assays, obtained by transfection of *E. coli* with the viral template DNA.

Mg^{2+} ion present in excess over the total dNTP concentration in the reaction. In these experiments we have not directly measured the amount of unbound metal ion present in the polymerase reactions. Therefore, since dNTPs, DNA, and proteins all bind Mg^{2+} ion, the absolute concentration of MgCl$_2$ which will give high fidelity DNA synthesis by *Taq* polymerase must be determined empirically for any given PCR protocol.

3.3.3 Reaction pH

The error rate of *Taq* polymerase can also be improved by decreasing the reaction pH (16). Low frequencies of both base substitution and frameshift errors (*Figure 5, Table 4*) are obtained when *Taq* polymerase reactions are carried out using Tris, 4-morpholine-ethane-sulphonic acid (Mes), and 1.4-piperazinebis(ethane-sulphonic acid) (Pipes) buffers at pH 5–6 (70 °C). The frequency of mutations continues to increase as the pH is raised through pH 8.2. The base substitution error rate of *Taq* polymerase changes approximately 60-fold and the frameshift error rate approximately 11-fold in response to a three-unit pH change (*Table 4*).

3.3.4 Temperature

The fidelity of the *Taq* polymerase has also been examined over a 50 °C temperature range (*Figure 6*). The error rate for base-substitutions (*Figure 6A*)

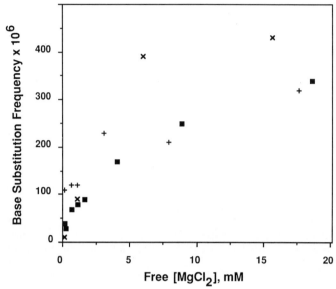

Figure 4. Fidelity of *Taq* polymerase versus free magnesium chloride concentration. The base-substitution mutant frequency was measured as described in *Figure 3*, except that the [dNTP] was varied along with the [MgCl$_2$]. Reversion frequencies are plotted as a function of 'Free [MgCl$_2$], which is the total MgCl$_2$ concentration added to the reactions minus the total added dNTP concentration. (■), 250 μM each dNTP (1 mM total); (+), 500 μM each dNTP (2 mM total); (×), 1 mM each dNTP (4 mM total).

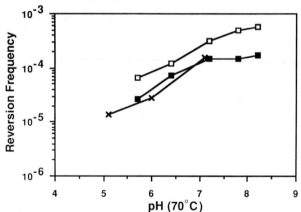

Figure 5. *Taq* polymerase fidelity versus reaction pH. The M13mp2 reversion frequencies of *Taq* polymerase reactions were measured as a function of pH using 20 mM Tris (□) or 20 mM Mes (×) buffers in the base-substitution assay, and 20 mM Tris buffer (■) in the frameshift assay. DNA synthesis reactions were performed using 50 mM KCl, 10 mM MgCl$_2$, and 1 mM each dNTP, allowing the buffers to equilibrate to 70°C for 10 min prior to addition of the remaining reagents (16). Reactions performed at pH<5.5 were incubated for 30 min to allow complete gap-filling synthesis. The plotted pH values are the average of the measured and calculated pH values at 70°C of mock reactions. (The differences between the measured and calculated pH values ranged from 0.31–0.50 pH units.)

Table 4. *Taq* DNA polymerase error rates under various conditions

DNA substrate	High fidelity conditions				Low fidelity conditions				Relative increase in fidelity[b]
	[MgCl$_2$], (mM)	[dNTP], (µM)[a]	pH	Error rate	[MgCl$_2$], (mM)	[dNTP] (µM)[a]	pH	Error rate	
A. *Base substitutions*									
A89 Opal codon	1	250	7.2	1/50 000	20	250	7.2	1/5400	9
	4	1000	7.2	1/300 000	20	1000	7.2	1/4200	71
	10	1000	5.1[c]	1/180 000	10	1000	8.2[d]	1/3200	56
Wild-type M13mp2	**1**	**250**	**7.2**	**1/58 000**	**10**	**1000**	**7.2**	**1/7500[e]**	**8**
B. *-1 base frameshifts*									
+T70 Frameshift[f]	2	500	7.2	<1/210 000	10	500	7.2	1/21 000	>10
	4	1000	7.2	<1/210 000	10	1000	7.2	1/24 000	≥9
	10	1000	5.7[d]	<1/210 000	10	1000	8.2[d]	1/19 000	>11
Wild-type M13mp2	**1**	**250**	**7.2**	**1/1 800 000**	**10**	**1000**	**7.2**	**1/35 000[e]**	**51**

DNA synthesis reactions were performed at 70°C as described in the legends to *Figures 1* and *2* in ref. 16, varying individual reaction components as indicated by the bold-face type.

[a] The total deoxynucleotide triphosphate concentration is four times the indicated value.
[b] High fidelity error rate ÷ Low fidelity error rate.
[c] Mes buffer
[d] Tris buffer
[e] Error rates were calculated from the mutant frequencies reported in ref. 13.
[f] Error rates were calculated assuming all mutations were at the run of five consecutive thymine bases.

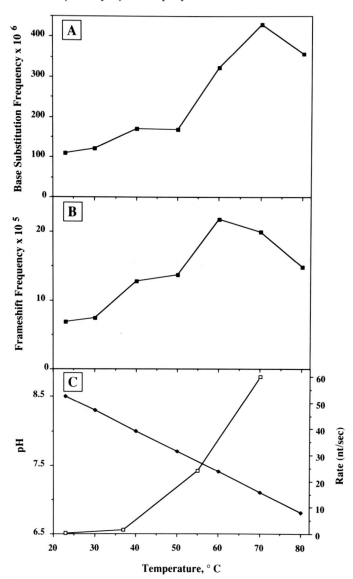

Figure 6. *Taq* polymerase fidelity versus reaction temperature. (**A**) Base-substitution frequency; (**B**) Frameshift frequency. Values plotted in (**A**) and (**B**) are the average of two experiments. (**C**) Change in reaction pH (◆) and polymerase synthetic rate (□) (17) as a function of temperature. Each polymerase reaction contained 20 mM Tris–HCl (pH 8.5 at room temp.), 50 mM KCl, 10 mM MgCl$_2$, 1 mM each dNTP, and a 1:1 molar equivalent (2.8 units of enzyme: 500 ng of gapped DNA) of *Taq* polymerase. Reaction mixtures were prewarmed to the indicated temperature for five minutes, after which time the enzyme was added. The length of incubation time was adjusted for the variation in rate with temperature to allow complete gap-filling synthesis.

and frameshifts (*Figure 6B*) increases several-fold as the reaction temperature is increased from 23–30 °C to 70–80 °C. The actual effect of temperature on fidelity could be underestimated since over the 50 °C temperature range the pH of the Tris buffer used in these reactions decreases by more than two units (*Figure 6C*), and fidelity is higher at low pH (*Figure 5*). *Figure 6C* also shows the variation in synthetic rate of the *Taq* polymerase with temperature (17). High fidelity PCR amplifications would thus be best performed using buffers such as Bis-Tris propane or Pipes which have a pK_a between pH 6 and pH 7 and a small temperature coefficient ($\Delta pK_a / °C$).

Incubation of DNA at high temperatures and low pH produces DNA damage, thus increasing the potential for mutations. One of the most frequent forms of DNA damage is deamination of cytosine to produce uracil (18). Since uracil has the same coding potential as thymine, faithful replication of a template uracil by a DNA polymerase will result in a $C \cdot G \to T \cdot A$ transition mutation. In PCR, the greatest risk of cytosine deamination is during the denaturation step. The rate constant for deamination of cytosine in single-stranded DNA at 80 °C is $\sim 1 \times 10^{-8}$/sec. This value increases to $\sim 2 \times 10^{-7}$/sec at 95 °C (*Table 5*).

A second common type of DNA damage is spontaneous base release resulting from hydrolysis of the N-glycosylic bond (18). The rate of this hydrolytic reaction is significantly increased at high temperatures and low pH. Depurination of native DNA at pH 7.4 and 70 °C occurs at a rate of $\sim 4 \times 10^{-9}$ per sec, while the rate of depyrimidination is twenty times slower ($\sim 2 \times 10^{-10}$/ sec at 80 °C). The release of pyrimidines, however, increases significantly as the temperature is increased to 95 °C (*Table 5*). Similarly, the rate of depurination (pH 5.0) increases two orders of magnitude as the temperature increases from 45 °C to 80 °C. Finally, lowering the pH markedly increases the rate of spontaneous purine loss in native DNA (*Table 5*). Following spontaneous base hydrolysis, the resulting apurinic/apyrimidinic (AP) sites in the DNA can inhibit synthesis by DNA polymerases (19). In some cases, DNA polymerases are capable of replicating past AP sites. As AP sites are non-coding lesions, such bypass replication is error-prone and results in base substitution mutations, most frequently transversions (19).

In conclusion, high temperature PCR amplification increases the likelihood of both *Taq* polymerase errors (*Figure 6*) and DNA damage-dependent errors (*Table 5*). To the extent possible, one should minimize PCR incubation times at elevated temperatures, especially when accurate DNA synthesis is required. For pH changes, the situation is more complex since *Taq* polymerase fidelity is higher at low pH (*Figure 5*), but low pH also increases the probability of DNA damage errors (*Table 5*). Varying the pH will thus provide a trade-off between improved polymerase fidelity and damage-induced infidelity.

3.3.5 Mutational specificity of *Taq* polymerase

We have previously described (13) the error specificity of *Taq* polymerase in

Table 5. Rate constants for cytosine deamination, depyrimidination, and depurination of DNA

Event	Temp. (°C)	pH	DNA form[a]	Rate constant (sec⁻¹)	Ref.
I. Deamination					
	37	7.4	ss	$1-2 \times 10^{-10}$	28, 29
			ds	7×10^{-13}	29
	50	7.4	ss	1×10^{-9}	29
	70	7.4	denatured, ss	$0.1-1 \times 10^{-8}$	28, 29
			ds	$<1 \times 10^{-10}$	28
	80	7.4	denatured, ss	$1-4 \times 10^{-8}$	28, 29
			ds	$<4 \times 10^{-10}$	28
	95	7.4	ss	2×10^{-7}	28
II. Depyrimidination					
	80	7.4	denatured	1×10^{-9}	30
			ds	$2-5 \times 10^{-10}$	30
	95	7.4	ss	2×10^{-8}	30
III. Depurination					
	37	7.4	ds	3×10^{-11}	31
	70	7.4	ds	4×10^{-9}	31
	45	5.0	ds	1×10^{-8}	31[b]
	60	5.0	ds	1×10^{-7}	31[b]
	80	5.0	ds	1×10^{-6}	31[b]
	70	5.0	ds	4×10^{-7}	31[c]
	70	6.0	ds	4×10^{-8}	31[c]
	70	7.0	ds	1×10^{-8}	31[c]

[a] ss, single-stranded; ds, double-stranded; denatured, heat-treated double-stranded
[b] Data taken from Figure 8 in ref. 31.
[c] Data taken from Figure 9 in ref. 31.

reactions containing 10 mM $MgCl_2$ and 4 mM total dNTPs, i.e. when the free Mg^{2+} concentration was 6 mM. Given the substantial improvement in fidelity provided by a low 'free' magnesium concentration (*Figure 3*), we determined the specificity of *Taq* polymerase errors from a reaction containing equal concentrations of both, i.e. 1 mM $MgCl_2$ and 1 mM total dNTPs. *Figure 7* shows the spectra of single-base mutations produced by *Taq* polymerase in the M13mp2 forward mutation assay under both conditions. The data are translated into quantitative base-substitution and frameshift error rates in *Table 4*. The effect of lowering the mutation frequency by decreasing the $MgCl_2$ concentration appears to be due to an equal decrease in all types of polymerase errors. The majority (70%) of single-base-substitution errors made by *Taq* polymerase under both reaction conditions are T→C transitions,

```
                                                    C
                                                    C
                                                    C
                                                  CCC
A                          A                      CCC  G
TGTGAGTTAG CTCACTCATT AGGCACCCCA GGCTTTACAC TTTATGCTTC CGGCTCGTAT GTTGTGTGGA ATTGTGAGCG GATAACAATT
C   -60    CT         -40   C              C    T    -20                      CGC  C  +1                  +20
C          C                                                                  CG   C


                                                              Δ
                                                              Δ                              A Δ
                              C            T                  Δ
TCACACAGGA AACAGCTATG      ACC ATG ATT ACG AAT TCA CTG GCC GTC GTT TTA CAA CGT CGT GAC TGG GAA AAC CCT
           +40                                     +60   Δ          C  +80  T              C
                                                                                           C

                                                                    T
                                                                    T
                                                                    C
                                                                    C
Δ                   C      C                     TA  Δ C   TA  Δ          Δ
GGC GTT ACC CAA CTT AAT CGC CTT GCA GCA CAT CCC CCT TTC GCC AGC TGG CGT AAT AGC GAA GAG GCC CGC ACC
G              C  +120 C                    T   +140             C  +160
                   C                                             C
```

Figure 7. Spectrum of single-base errors produced by *Taq* polymerase. Three lines of primary DNA sequence are shown. The sequence is that of the viral (+) template strand. Position +1 is the first transcribed base. The spectrum above the lines of primary DNA sequence is composed of mutants from low fidelity reactions using 10 mM MgCl$_2$, 1 mM dNTPs, pH 7.2 (70°C). The spectrum below the lines of primary DNA sequence is mutants obtained from high fidelity reactions: 1 mM MgCl$_2$, 250 μM dNTPs, pH 7.2 (70°C). Base-substitutions are indicated by letters for the new base found in the viral DNA. For frameshifts, the loss of a base is indicated by an open triangle, while the addition of a base is indicated by a filled triangle. When frameshifts occur at reiterated nucleotide positions, it is not possible to distinguish which base was lost or added, and the symbol is centred within the run.

which arise through a T·dGTP mispair intermediate. *Taq* polymerase also produces one-base deletion errors, albeit at a much lower frequency than for base-substitutions.

4. Determining fidelity for a given PCR application

The overall DNA polymerase fidelity that is needed during a given PCR experiment will vary depending on the precise nature of the PCR application. For some uses of PCR, such as random mutagenesis of a DNA sequence, a high polymerase error rate would be desirable. For those studies which use the final PCR products in a biochemical procedure such as DNA sequencing or filter hybridization studies, random polymerase errors usually can be tolerated, since the proportion of modified DNA moleclues present in the population is below the detection limit of the assay. A mathematical discussion of the confidence levels for such analyses has recently been published (20). However, several examples of PCR techniques currently exist in the literature which require that a high level of fidelity be maintained throughout the amplification procedure. Although varied in aim, such techniques share the common feature of having a genetically heterogeneous amplified population of DNA molecules. Examples include the detection of mutant or variant genes that occur with a relatively low frequency (10, 21), the construction of genetically engineered monoclonal antibodies (22), the analysis of T-cell receptor allelic polymorphism and the generation of T-cell receptor diversity (23, 24), the study of HIV variation *in vivo* (25), and the study of meiotic recombination in individual cells (26). Finally, cases will also exist wherein the investigation requires the cloning of individual DNA molecules from the amplified population, in which case a low polymerase error rate per cycle of amplification will minimize the number of unwanted changes in the cloned molecules.

The proportion of altered DNA molecules which will be present in the population after PCR amplification is a function of the DNA polymerase error rate per cycle, the number of amplification cycles and the starting population size. To a first approximation, the occurrence of polymerase errors during PCR-DNA amplification follows a Poisson distribution for rare and random events, similar to the occurrence of spontaneous mutations during the growth of a bacterial culture (27). Assuming 100% amplification efficiency with each cycle, the expected number of polymerase errors at the end of a PCR experiment, $E(x)$, is the product of the chance of error per cycle ($s_0 p$) and the increase in the total number of DNA molecules ($n2^{n-1}$), where s_0 is the starting population size in nucleotides, p is the polymerase error rate per nucleotide synthesized per cycle, and n is the number of amplification cycles. The error frequency, f, defined as the proportion of expected nucleotide errors in the final population after n cycles is thus,

$$f = \frac{(s_0 p)\,(n2^{n-1})}{s_0 2^n}, \text{ or } \frac{np}{2}.$$

The population of DNA molecules containing errors increases during PCR from two sources: newly arising polymerase errors at each cycle, and the amplification of DNA molecules containing errors from previous cycles. The above formula accurately describes the average error frequency for PCR amplifications in which the starting population size s_0 is equal to or greater than $1/p$. On average, the same number of DNA sequence changes in the final population result from polymerase errors during the last amplification cycle as result from the amplification of errors which occurred during previous cycles. However, the rare occurrence of an early error can increase the final error frequency above the average described by $f = np/2$. The distribution which describes the occurrence of mutations in a population is characterized as having an abnormally high variance, i.e. a significant number of populations with a larger than average number of mutants (27). This characteristic can be compared to the operation of a slot machine (27). The average monetary return from a limited number of plays is smaller than the initial amount of money used in play; however, when the improbable 'jackpot' is hit, the amount of money returned is far greater than the input amount. Therefore, in any amplification reaction, there is some probability that an error will occur during the first few cycles of DNA synthesis, resulting in a 'jackpot' reaction wherein the population error frequency exceeds that predicted by $f = np/2$.

A tabulation of expected average error frequences when $s_0 \geq p$ is given in *Table 6*. As shown here, one can experimentally manipulate the probability of DNA sequence changes by altering the number of cycles (n) and/or the polymerase error rate per nucleotide (p). For example, the expected error frequency when $p = 1/10\,000$ is 1×10^{-3} after 20 cycles, or one error per 1000 nucleotides. This error frequency increases to 2.5×10^{-3} or one error per 400 nucleotides after 50 amplification cycles. However, after 50 cycles when $p = 1/50\,000$, the error frequency is 5×10^{-4} (1/2000); when $p = 1/200\,000$, the expected number of errors is only one error in 8000 nucleotides. This descrip-

Table 6. Expected error frequencies during PCR amplification

	Error frequency $\times$ 100 ($f = np/2$)		
n	$p = 1/10\,000$	$p = 1/50\,000$	$p = 1/200\,000$
1	0.005	0.001	0.00025
2	0.010	0.002	0.0005
5	0.025	0.05	0.00125
10	0.05	0.01	0.0025
20	0.10	0.02	0.005
50	0.25	0.05	0.0125

tion of error frequency does not address the question of how many changes an individual DNA molecule is expected to contain, but only estimates the total number of changes in the final DNA population.

5. Recommendations for high fidelity PCR

Before performing high fidelity PCR, one should first evaluate the final error frequency which can be tolerated by a particular PCR application. Giving sufficient attention to the details of a PCR reaction should enable researchers to effectively use PCR products for a variety of applications, including those that require the stringent maintenance of the original DNA sequence content. Several experimental alternatives exist to achieve this goal. Parameters to consider include:

(a) **Cycle number**. Keep the number of cycles to the minimum required to produce a feasible amount of product DNA. Remember, $f = np/2$, and many molecular biology manipulations can be carried out with a picomole or less DNA.

(b) **Polymerase**. When technically possible, employ a DNA polymerase having a highly active proofreading exonuclease, such as the T4 or native T7 polymerase, to synthesize DNA. These polymerases have a much higher fidelity than the exonuclease-deficient, thermostable polymerases.

(c) **Reaction conditions**. When using *Taq* polymerase one should realize that conditions that may yield higher fidelity may not necessarily be optimal for PCR product yield. With this in mind:

- Keep the four deoxynucleotide triphosphate precursors at equal concentrations, and the total dNTP concentration at the lowest necessary to support the desired amount of DNA synthesis.

- Keep reaction times as short as is necessary to support the desired amount of synthesis, as this will minimize the time available for the polymerase to extend from mispaired termini.

- The $MgCl_2$ concentration should be as low as can support the desired amount of synthesis and should not be in large excess over the total dNTP concentration.

- Minimize reaction time at high temperatures, since polymerase fidelity will be higher and less DNA damage will occur.

- For a small number of cycles, the fidelity of amplification may be limited more by *Taq* polymerase errors with undamaged substrates than by DNA damage produced by the incubation conditions characteristic of PCR. Thus, in the pH trade-off between higher polymerase fidelity and increased DNA damage, higher fidelity may be obtained by performing reactions at ~pH 6 (70°C), somewhat lower than the pH generally used for PCR.

(d) **Verify the DNA sequence**. The DNA sequence of independent clones from the final PCR population should be determined to ensure that the original DNA sequence has been maintained during the amplification process. Ideally, the clones should be derived from separate PCR experiments to minimize the risk of over-representation of early polymerase errors in 'jackpot' reactions.

References

1. Kornberg, A. (1980). *DNA Replication*. Freeman, San Francisco, Calif.
2. Kuchta, R. D., Benkovic, P., and Benkovic, S. J. (1988). *Biochemistry*, **27**, 6716.
3. Mendelman, L. V., Boosalis, M. S., Petruska, J., and Goodman, M. F. (1989). *Journal of Biological Chemistry*, **264**, 14415.
4. Bebenek, K., Joyce, C. M., Fitzgerald, M. P., and Kunkel, T. A. (1990). *Journal of Biological Chemistry*. **265**, 13878.
5. Kunkel, T. A. and Bebenek, K. (1988). *Biochimica et Biophysica Acta*, **951**, 1.
6. Goodman, M. F. (1988). *Mutation Research*, **200**, 11.
7. Mendelman, L. V., Petruska, J., and Goodman, M. F. (1990). *Journal of Biological Chemistry*, **265**, 2338.
8. Kunkel, T. A. (1988). *Cell*, **53**, 837.
9. Keohavong, P. and Thilly, W. G. (1989). *Proceedings of the National Academy of Sciences, USA*, **86**, 9253.
10. Loeb, L. A. and Kunkel, T. A. (1982). *Annual Reviews in Biochemistry*, **52**, 429.
11. Kunkel, T. A. (1985). *Journal of Biological Chemistry*, **260**, 5787.
12. Kunkel, T. A. and Soni, A. (1988). *Journal of Biological Chemistry*, **263**, 4450.
13. Tindall, K. R. and Kunkel, T. A. (1988). *Biochemistry*, **27**, 6008.
14. Saiki, R. K., Gelfand, D. H., Stoffel, S., Scharf, S. J., Higuchi, R., Horn, G. T., Mullis, K. B., and Erlich, H. A. (1988). *Science*, **239**, 487.
15. Erlich, H. A., Gibbs, R., and Kazazian, H. H. (ed.) (1989). *Current communications in molecular biology: Polymerase chain reaction*. Cold Spring Harbor Laboratory Press, Cold Spring Harbor, NY.
16. Eckert, K. A. and T. A. Kunkel (1990) *Nucleic Acids Research*, **18**, 3729.
17. Innis, M. A., Myambo, K. B., Gelfand, D. H., and Brow, M. D. (1988). *Proceedings of the National Academy of Sciences, USA*, **85**, 9436.
18. Lindahl, T. (1979). *Progress in Nucleic Acids Research and Molecular Biology*, **22**, 135.
19. Loeb, L. A. and Preston, B. D. (1986). *Annual Reviews in Genetics*, **20**, 201.
20. Krawczak, M., Reiss, J., and Rosler, U. (1989). *Nucleic Acids Research*, **17**, 2197.
21. Vrieling, H., Van Roooijen, M. L., Groen, N. A., Zdzienicka, M. Z., Simons, J. W., Lohman, P. H., and Van Zeeland, A. A. (1989). *Molecular and Cell Biology*, **9**, 1277.
22. Larrick, J. W., Danielsson, L., Brenner, C. A., Wallace, E. F., Abrahamson, M., Fry, K. E., and Borrebaeck, C. A. (1989). *Biotechnology*, **7**, 934.
23. Loh, E. Y., Elliott, J. F., Cwirla, S., Lanier, L. L., and Davis, M. M. (1989). *Science*, **243**, 217.
24. Lacy, M. J., McNeil, L. K., Roth, M. E., and Kranz, D. M. (1989) *Proceedings of the National Academy of Sciences, USA*, **86**, 1023.

25. Goodenow, M., Juet, T., Saurin, W., Kwok, S., Sninsky, J., and Wain-Hobson, S. (1989), *JAIDS*, **2**, 344.
26. Li, H., Gyllensten, U. B., Cui, X., Saiki, R. K., Erlich, H. A., and Arnheim, N. (1988). *Nature*, **335,** 414.
27. Luria, S. E. and Delbruck, M. (1943). *Genetics*, **28,** 491.
28. Lindahl, T. and Nyberg, B. (1974) *Biochemistry*, **13,** 3405.
29. Frederico, L. A., Kunkel, T. A., and Shaw, B. (1990). *Biochemistry*, **29**, 2532.
30. Lindahl, T. and Karlstrom, O. (1973). *Biochemistry*, **12**, 5151.
31. Lindahl, T. and Nyberg, B. (1972). *Biochemistry*, **11**, 3610.

Appendix

Suppliers of specialist items

Aldrich Chemical Co., 940 West Saint Paul Avenue, Milwaukee, WI 53223, USA; The Old Brickyard, New Road, Gillingham, Dorset, SP8 4JL, UK.

Amersham International plc, White Lion Road, Amersham, Bucks HP7 9LL, UK; 2636 S. Clearbrook Drive, Arlington Heights, Il 60005, USA.

Amicon, Upper Mill, Stonehouse, Gloucester, GL10 2BJ, UK; 17 Cherry Hill Drive, Danvers, MA 01923, USA.

Anachem Ltd., Anachem House, 10 Charles Street, Luton, Beds, LU2 0EB, UK.

Anderman & Company Ltd, 145 London Road, Kingston upon Thames, Surrey KT2 6NH, UK.

Anglian Biotechnology, Whitehall House, Whitehall Road, Colchester, Essex CO2 8HA, UK.

Applied Biosystems Inc., 850 Lincoln Centre Drive, Foster City, CA 94404, USA. Biotech Instruments Ltd., Unit A, Caxton Hill Extension Road, Caxton Hill, Hertford, SG13 7LS, UK.

Beckman Instruments Inc., 2500 Harbour Boulevard, P.O. Box 3100, Fullerton, CA 92634, USA.

Bio 101 Inc, PO Box 2284, La Jolla, Ca 92038–2284, USA; Stratech Ltd., 61/63 Dudley Street, Luton, Beds, LU2 0NP, UK.

Bio-Rad Laboratories, 1414, Harbor Way South, Richmond, CA 94804, USA; Caxton Way, Watford Business Park, Watford, Herts, WD1 8RP, UK.

Biomed Instruments Inc., 1020 South Raymound Avenue, Suite B, Fullerton, CA 92631, USA.

Biometra, P.O. Box 167, Maidstone, Kent, ME14 2AT, UK.

Biotherm Corporation, 3260 Wilson Blvd, Arlington, VA 22201, USA.

Boehringer-Mannheim GmbH, Postfach 310120, D-6800, Mannhem 31, West Germany; P.O. Box 50816, Indianapolis, IN 46250, USA.

Brinkmann Instruments, Cantiague Road, Westbury, NY 11590, USA.

Cambio, 34 Millington Road, Newnham, Cambridge, CB3 9HP, UK.

Cetus, *see* Perkin Elmer Cetus.

Cole Parmer Instrument Co., 7425 North Oak Park Avenue, Chicago, Illinois

60648, USA. CP Laboratories, P.O. Box 22, Bishop's Stortford, Herts, CM23 3DX, UK.

Coy Laboratory Products, 22 Metty Dr., Ann Arbor, MI 48103, USA; Flowgen Ltd., Broad Oak Enterprise Village, Broad Lane, Sittingbourne, Kent, ME9 8AQ, UK.

Dynal Ltd., Station House, 26 Grove Street, New Ferry, Wirral, L62 5AZ, UK.

DuPont Company, Biotechnology Systems Division, BRML, G-50986, Wilmington, DE 19898, USA; Wedgewood Way, Stevenage, Herts, SG1 4QN, UK.

Eastman Kodak, Acorn Field Road, Knowsley Industrial Park North, Liverpool, L33 72X, UK; Rochester, New York 14650, USA.

Eppendorf Gerätebau Netheler and Hinz GmbH, P.O. Box 650670, D-2000 Hamburg, West Germany.

Fluka, Peakdale Road, Glossop, Derby, SK13 9XE, UK.

FMC Bioproducts, Rockland, Maine, USA. Flowgen Instruments Ltd., Broad Oak Enterprise Village, Broad Road, Sittingbourne, Kent, ME9 8AQ, UK.

Gibco BRL Life Technologies Inc., Grand Island, NT, USA; P.O. Box 35, Trident House, Renfrew Road, Paisley, PA3 4EF, UK.

Gilson France SA, Box 27, 3000 West Beltline Hwy., Middleton, WI 53562, USA; 72 rue Gambetta, BP 45, F-95400 Villiers-le-Biel, France.

Grant Instruments (Cambridge) Ltd, Barrington, Cambridge CB2 5QZ, UK.

Hoefer Scientific Instruments, 654 Minnesota Street, P.O. Box 77387, San Francisco, CA 94107, USA; Unit 12, Croft Road Workshops, Croft Road, Newcastle under Lyme, ST5 0TH, UK.

Hybaid, 111-113 Waldegrave Road, Teddington, Middlesex, TW11 8LL, UK.

IBI Ltd., 36 Clifton Road, Cambridge, CB1 4ZR, UK; P.O. Box 9558, New Haven, CT 06535, USA.

ICN Biochemicals, PO Box 28050, Cleveland, DH 44128, USA; Eagle House, Peregrine Business Park, Gomm Road, High Wycombe, Bucks, HP13 7DL, UK.

ILS Ltd., 14-15 Newbury Street, London EC1A 7HU, UK (A division of Perkin Elmer Ltd).

Kinematica UK, Philip Harris Scientific, 618 Western Avenue, Park Royal, London W3 0TE, UK.

Koch Light, NBS Biologicals, Edison House, 163 Dixons Hill Road, Hatfield, AL9 7JE, UK; NBS Biologicals, Eddison, New Jersey, USA.

LEP Scientific, Sunrise Parkway, Linford Wood (East), Milton Keynes, Bucks, MK14 6QF, UK.

MJ Research, Kendall Square Box 363, Cambridge, MA 02142, USA; Genetic Research Instrumentation, Gene House, Dunmow Road, Felstead, Dunmow, Essex CM6 3LD, UK.

New England Biolabs, 32 Tozer Road, Beverly, MA 01915-9990, USA; Postach 2750, 6231 Schwalbach/Taunus, West Germany; CP Laboratories, P.O. Box 22, Bishop's Stortford, Herts, CM23 3DX, UK.

New England Nuclear/DuPont, Barley Hill Plaza, P24–2174, Wilmington, DE 19898, USA; Wedgwood Way, Stevenage, Herts, SG1 6Y1, UK.

Perkin Elmer Cetus, Main Avenue, Norwalk, CT 06856, USA; Maxwell Road, Beaconsfield, Bucks, HP9 1QA, UK.

Pharmacia-LKB Biotechnology AB, S-75182, Uppsala, Sweden; 800 Centennial Avenue, Piscataway, NJ 08854, USA.

Promega Corporation, 2800 South Fish Hatchery Road, Madison, WI 53711–5305, USA; Episcon House, Enterprise Road, Chilworth Research Centre, Southampton, SO1 7NS, UK.

Research Genetics Inc., Huntsville, Alabama, USA.

Science/Electronics, PO Box 386, Dayton, OH 45401, USA.

Sigma Chemical Company Ltd, Fancy Road, Poole, Dorset BH17 7NH, UK.

Stratagene Ltd., Cambridge Innovation Centre, Cambridge Sciences Park, Milton Road, Cambridge, CB4 4GF, UK; 11099 North Torrey Pines Road, La Jolla, CA 92037, USA.

VA Howe Ltd., 12-14 St. Ann's Crescent, London, SW18 2LS, UK.

Whatman Lab Sales Ltd., Unit 1, Coldred Road, Parkwood, Maidstone, Kent, ME15 9XN, UK.

Index

Index

Index

Index

Index